Performance Assessment of Concrete Structures and Engineered Barriers for Nuclear Applications

RILEM STATE-OF-THE-ART REPORTS

Volume 21

RILEM, The International Union of Laboratories and Experts in Construction Materials, Systems and Structures, founded in 1947, is a non-governmental scientific association whose goal is to contribute to progress in the construction sciences, techniques and industries, essentially by means of the communication it fosters between research and practice. RILEM's focus is on construction materials and their use in building and civil engineering structures, covering all phases of the building process from manufacture to use and recycling of materials. More information on RILEM and its previous publications can be found on www.RILEM.net.

The RILEM State-of-the-Art Reports (STAR) are produced by the Technical Committees. They represent one of the most important outputs that RILEM generates—high level scientific and engineering reports that provide cutting edge knowledge in a given field. The work of the TCs is one of RILEM's key functions. Members of a TC are experts in their field and give their time freely to share their expertise. As a result, the broader scientific community benefits greatly from RILEM's activities.

RILEM's stated objective is to disseminate this information as widely as possible to the scientific community. RILEM therefore considers the STAR reports of its TCs as of highest importance, and encourages their publication whenever possible.

The information in this and similar reports is mostly pre-normative in the sense that it provides the underlying scientific fundamentals on which standards and codes of practice are based. Without such a solid scientific basis, construction practice will be less than efficient or economical.

It is RILEM's hope that this information will be of wide use to the scientific community.

More information about this series at http://www.springer.com/series/8780

Valérie L'Hostis · Robert Gens
Editors

Performance Assessment of Concrete Structures and Engineered Barriers for Nuclear Applications

Conclusions of RILEM TC 226-CNM

Springer

Editors
Valérie L'Hostis
Laboratory of 'Etude du comportement des bétons et des argiles'
CEA Saclay
Gif-sur-Yvette
France

Robert Gens
ONDRAF-NIRAS
Brussels
Belgium

ISSN 2213-204X ISSN 2213-2031 (electronic)
RILEM State-of-the-Art Reports
ISBN 978-94-024-0903-1 ISBN 978-94-024-0904-8 (eBook)
DOI 10.1007/978-94-024-0904-8

Library of Congress Control Number: 2016946004

Printed on acid-free paper

This Springer imprint is published by Springer Nature
The registered company is Springer Science+Business Media B.V. Dordrecht

RILEM Publications

The following list is presenting the global offer of RILEM Publications, sorted by series. Each publication is available in printed version and/or in online version.

RILEM Proceedings (PRO)

PRO 1: Durability of High Performance Concrete (ISBN: 2-912143-03-9); *Ed. H. Sommer*

PRO 2: Chloride Penetration into Concrete (ISBN: 2-912143-00-04); *Eds. L.-O. Nilsson and J.-P. Ollivier*

PRO 3: Evaluation and Strengthening of Existing Masonry Structures (ISBN: 2-912143-02-0); *Eds. L. Binda and C. Modena*

PRO 4: Concrete: From Material to Structure (ISBN: 2-912143-04-7); *Eds. J.-P. Bournazel and Y. Malier*

PRO 5: The Role of Admixtures in High Performance Concrete (ISBN: 2-912143-05-5); *Eds. J. G. Cabrera and R. Rivera-Villarreal*

PRO 6: High Performance Fiber Reinforced Cement Composites—HPFRCC 3 (ISBN: 2-912143-06-3); *Eds. H. W. Reinhardt and A. E. Naaman*

PRO 7: 1st International RILEM Symposium on Self-Compacting Concrete (ISBN: 2-912143-09-8); *Eds. Å. Skarendahl and Ö. Petersson*

PRO 8: International RILEM Symposium on Timber Engineering (ISBN: 2-912143-10-1); *Ed. L. Boström*

PRO 9: 2nd International RILEM Symposium on Adhesion between Polymers and Concrete ISAP '99 (ISBN: 2-912143-11-X); *Eds. Y. Ohama and M. Puterman*

PRO 10: 3rd International RILEM Symposium on Durability of Building and Construction Sealants (ISBN: 2-912143-13-6); *Ed. A. T. Wolf*

PRO 11: 4th International RILEM Conference on Reflective Cracking in Pavements (ISBN: 2-912143-14-4); *Eds. A. O. Abd El Halim, D. A. Taylor and El H. H. Mohamed*

PRO 12: International RILEM Workshop on Historic Mortars: Characteristics and Tests (ISBN: 2-912143-15-2); *Eds. P. Bartos, C. Groot and J. J. Hughes*

PRO 13: 2nd International RILEM Symposium on Hydration and Setting (ISBN: 2-912143-16-0); *Ed. A. Nonat*

PRO 14: Integrated Life-Cycle Design of Materials and Structures—ILCDES 2000 (ISBN: 951-758-408-3); (ISSN: 0356-9403); *Ed. S. Sarja*

PRO 15: Fifth RILEM Symposium on Fibre-Reinforced Concretes (FRC)—BEFIB'2000 (ISBN: 2-912143-18-7); *Eds. P. Rossi and G. Chanvillard*

PRO 16: Life Prediction and Management of Concrete Structures (ISBN: 2-912143-19-5); *Ed. D. Naus*

PRO 17: Shrinkage of Concrete—Shrinkage 2000 (ISBN: 2-912143-20-9); *Eds. V. Baroghel-Bouny and P.-C. Aïtcin*

PRO 18: Measurement and Interpretation of the On-Site Corrosion Rate (ISBN: 2-912143-21-7); *Eds. C. Andrade, C. Alonso, J. Fullea, J. Polimon and J. Rodriguez*

PRO 19: Testing and Modelling the Chloride Ingress into Concrete (ISBN: 2-912143-22-5); *Eds. C. Andrade and J. Kropp*

PRO 20: 1st International RILEM Workshop on Microbial Impacts on Building Materials (CD 02) (e-ISBN 978-2-35158-013-4); *Ed. M. Ribas Silva*

PRO 21: International RILEM Symposium on Connections between Steel and Concrete (ISBN: 2-912143-25-X); *Ed. R. Eligehausen*

PRO 22: International RILEM Symposium on Joints in Timber Structures (ISBN: 2-912143-28-4); *Eds. S. Aicher and H.-W. Reinhardt*

PRO 23: International RILEM Conference on Early Age Cracking in Cementitious Systems (ISBN: 2-912143-29-2); *Eds. K. Kovler and A. Bentur*

PRO 24: 2nd International RILEM Workshop on Frost Resistance of Concrete (ISBN: 2-912143-30-6); *Eds. M. J. Setzer, R. Auberg and H.-J. Keck*

PRO 25: International RILEM Workshop on Frost Damage in Concrete (ISBN: 2-912143-31-4); *Eds. D. J. Janssen, M. J. Setzer and M. B. Snyder*

PRO 26: International RILEM Workshop on On-Site Control and Evaluation of Masonry Structures (ISBN: 2-912143-34-9); *Eds. L. Binda and R. C. de Vekey*

PRO 27: International RILEM Symposium on Building Joint Sealants (CD03); *Ed. A. T. Wolf*

PRO 28: 6th International RILEM Symposium on Performance Testing and Evaluation of Bituminous Materials—PTEBM'03 (ISBN: 2-912143-35-7; e-ISBN: 978-2-912143-77-8); *Ed. M. N. Partl*

PRO 29: 2nd International RILEM Workshop on Life Prediction and Ageing Management of Concrete Structures (ISBN: 2-912143-36-5); *Ed. D. J. Naus*

PRO 30: 4th International RILEM Workshop on High Performance Fiber Reinforced Cement Composites—HPFRCC 4 (ISBN: 2-912143-37-3); *Eds. A. E. Naaman and H. W. Reinhardt*

PRO 31: International RILEM Workshop on Test and Design Methods for Steel Fibre Reinforced Concrete: Background and Experiences (ISBN: 2-912143-38-1); *Eds. B. Schnütgen and L. Vandewalle*

PRO 32: International Conference on Advances in Concrete and Structures 2 vol. (ISBN (set): 2-912143-41-1); *Eds. Ying-shu Yuan, Surendra P. Shah and Heng-lin Lü*

PRO 33: 3rd International Symposium on Self-Compacting Concrete (ISBN: 2-912143-42-X); *Eds. Ó. Wallevik and I. Níelsson*

PRO 34: International RILEM Conference on Microbial Impact on Building Materials (ISBN: 2-912143-43-8); *Ed. M. Ribas Silva*

PRO 35: International RILEM TC 186-ISA on Internal Sulfate Attack and Delayed Ettringite Formation (ISBN: 2-912143-44-6); *Eds. K. Scrivener and J. Skalny*

PRO 36: International RILEM Symposium on Concrete Science and Engineering—A Tribute to Arnon Bentur (ISBN: 2-912143-46-2); *Eds. K. Kovler, J. Marchand, S. Mindess and J. Weiss*

PRO 37: 5th International RILEM Conference on Cracking in Pavements—Mitigation, Risk Assessment and Prevention (ISBN: 2-912143-47-0); *Eds. C. Petit, I. Al-Qadi and A. Millien*

PRO 38: 3rd International RILEM Workshop on Testing and Modelling the Chloride Ingress into Concrete (ISBN: 2-912143-48-9); *Eds. C. Andrade and J. Kropp*

PRO 39: 6th International RILEM Symposium on Fibre-Reinforced Concretes—BEFIB 2004 (ISBN: 2-912143-51-9); *Eds. M. Di Prisco, R. Felicetti and G. A. Plizzari*

PRO 40: International RILEM Conference on the Use of Recycled Materials in Buildings and Structures (ISBN: 2-912143-52-7); *Eds. E. Vázquez, Ch. F. Hendriks and G. M. T. Janssen*

PRO 41: RILEM International Symposium on Environment-Conscious Materials and Systems for Sustainable Development (ISBN: 2-912143-55-1); *Eds. N. Kashino and Y. Ohama*

PRO 42: SCC'2005—China: 1st International Symposium on Design, Performance and Use of Self-Consolidating Concrete (ISBN: 2-912143-61-6); *Eds. Zhiwu Yu, Caijun Shi, Kamal Henri Khayat and Youjun Xie*

PRO 43: International RILEM Workshop on Bonded Concrete Overlays (e-ISBN: 2-912143-83-7); *Eds. J. L. Granju and J. Silfwerbrand*

PRO 44: 2nd International RILEM Workshop on Microbial Impacts on Building Materials (CD11) (e-ISBN: 2-912143-84-5); *Ed. M. Ribas Silva*

PRO 45: 2nd International Symposium on Nanotechnology in Construction, Bilbao (ISBN: 2-912143-87-X); *Eds. Peter J. M. Bartos, Yolanda de Miguel and Antonio Porro*

PRO 46: ConcreteLife'06—International RILEM-JCI Seminar on Concrete Durability and Service Life Planning: Curing, Crack Control, Performance in Harsh Environments (ISBN: 2-912143-89-6); *Ed. K. Kovler*

PRO 47: International RILEM Workshop on Performance Based Evaluation and Indicators for Concrete Durability (ISBN: 978-2-912143-95-2); *Eds. V. Baroghel-Bouny, C. Andrade, R. Torrent and K. Scrivener*

PRO 48: 1st International RILEM Symposium on Advances in Concrete through Science and Engineering (e-ISBN: 2-912143-92-6); *Eds. J. Weiss, K. Kovler, J. Marchand, and S. Mindess*

PRO 49: International RILEM Workshop on High Performance Fiber Reinforced Cementitious Composites in Structural Applications (ISBN: 2-912143-93-4); *Eds. G. Fischer and V.C. Li*

PRO 50: 1st International RILEM Symposium on Textile Reinforced Concrete (ISBN: 2-912143-97-7); *Eds. Josef Hegger, Wolfgang Brameshuber and Norbert Will*

PRO 51: 2nd International Symposium on Advances in Concrete through Science and Engineering (ISBN: 2-35158-003-6; e-ISBN: 2-35158-002-8); *Eds. J. Marchand, B. Bissonnette, R. Gagné, M. Jolin and F. Paradis*

PRO 52: Volume Changes of Hardening Concrete: Testing and Mitigation (ISBN: 2-35158-004-4; e-ISBN: 2-35158-005-2); *Eds. O. M. Jensen, P. Lura and K. Kovler*

PRO 53: High Performance Fiber Reinforced Cement Composites—HPFRCC5 (ISBN: 978-2-35158-046-2); *Eds. H. W. Reinhardt and A. E. Naaman*

PRO 54: 5th International RILEM Symposium on Self-Compacting Concrete (ISBN: 978-2-35158-047-9); *Eds. G. De Schutter and V. Boel*

PRO 55: International RILEM Symposium Photocatalysis, Environment and Construction Materials (ISBN: 978-2-35158-056-1); *Eds. P. Baglioni and L. Cassar*

PRO 56: International RILEM Workshop on Integral Service Life Modelling of Concrete Structures (ISBN 978-2-35158-058-5); *Eds. R. M. Ferreira, J. Gulikers and C. Andrade*

PRO 57: RILEM Workshop on Performance of cement-based materials in aggressive aqueous environments (e-ISBN: 978-2-35158-059-2); *Ed. N. De Belie*

PRO 58: International RILEM Symposium on Concrete Modelling—CONMOD'08 (ISBN: 978-2-35158-060-8); *Eds. E. Schlangen and G. De Schutter*

PRO 59: International RILEM Conference on On Site Assessment of Concrete, Masonry and Timber Structures—SACoMaTiS 2008 (ISBN set: 978-2-35158-061-5); *Eds. L. Binda, M. di Prisco and R. Felicetti*

PRO 60: Seventh RILEM International Symposium on Fibre Reinforced Concrete: Design and Applications—BEFIB 2008 (ISBN: 978-2-35158-064-6); *Ed. R. Gettu*

PRO 61: 1st International Conference on Microstructure Related Durability of Cementitious Composites 2 vol., (ISBN: 978-2-35158-065-3); *Eds. W. Sun, K. van Breugel, C. Miao, G. Ye and H. Chen*

PRO 62: NSF/ RILEM Workshop: In-situ Evaluation of Historic Wood and Masonry Structures (e-ISBN: 978-2-35158-068-4); *Eds. B. Kasal, R. Anthony and M. Drdácký*

PRO 63: Concrete in Aggressive Aqueous Environments: Performance, Testing and Modelling, 2 vol., (ISBN: 978-2-35158-071-4); *Eds. M. G. Alexander and A. Bertron*

PRO 64: Long Term Performance of Cementitious Barriers and Reinforced Concrete in Nuclear Power Plants and Waste Management—NUCPERF 2009 (ISBN: 978-2-35158-072-1); *Eds. V. L'Hostis, R. Gens, C. Gallé*

PRO 65: Design Performance and Use of Self-consolidating Concrete—SCC'2009 (ISBN: 978-2-35158-073-8); *Eds. C. Shi, Z. Yu, K. H. Khayat and P. Yan*

PRO 66: 2nd International RILEM Workshop on Concrete Durability and Service Life Planning—ConcreteLife'09 (ISBN: 978-2-35158-074-5); *Ed. K. Kovler*

PRO 67: Repairs Mortars for Historic Masonry (e-ISBN: 978-2-35158-083-7); *Ed. C. Groot*

PRO 68: Proceedings of the 3rd International RILEM Symposium on 'Rheology of Cement Suspensions such as Fresh Concrete (ISBN 978-2-35158-091-2); *Eds. O. H. Wallevik, S. Kubens and S. Oesterheld*

PRO 69: 3rd International PhD Student Workshop on 'Modelling the Durability of Reinforced Concrete' (ISBN: 978-2-35158-095-0); *Eds. R. M. Ferreira, J. Gulikers and C. Andrade*

PRO 70: 2nd International Conference on 'Service Life Design for Infrastructure' (ISBN set: 978-2-35158-096-7, e-ISBN: 978-2-35158-097-4); *Ed. K. van Breugel, G. Ye and Y. Yuan*

PRO 71: Advances in Civil Engineering Materials—The 50-year Teaching Anniversary of Prof. Sun Wei' (ISBN: 978-2-35158-098-1; e-ISBN: 978-2-35158-099-8); *Eds. C. Miao, G. Ye, and H. Chen*

PRO 72: First International Conference on 'Advances in Chemically-Activated Materials—CAM'2010' (2010), 264 pp., ISBN: 978-2-35158-101-8; e-ISBN: 978-2-35158-115-5, *Eds. Caijun Shi and Xiaodong Shen*

PRO 73: 2nd International Conference on 'Waste Engineering and Management—ICWEM 2010' (2010), 894 pp., ISBN: 978-2-35158-102-5; e-ISBN: 978-2-35158-103-2, *Eds. J. Zh. Xiao, Y. Zhang, M. S. Cheung and R. Chu*

PRO 74: International RILEM Conference on 'Use of Superabsorsorbent Polymers and Other New Addditives in Concrete' (2010) 374 pp., ISBN: 978-2-35158-104-9; e-ISBN: 978-2-35158-105-6; *Eds. O.M. Jensen, M.T. Hasholt, and S. Laustsen*

PRO 75: International Conference on 'Material Science—2nd ICTRC—Textile Reinforced Concrete—Theme 1' (2010) 436 pp., ISBN: 978-2-35158-106-3; e-ISBN: 978-2-35158-107-0; *Ed. W. Brameshuber*

PRO 76: International Conference on 'Material Science—HetMat—Modelling of Heterogeneous Materials—Theme 2' (2010) 255 pp., ISBN: 978-2-35158-108-7; e-ISBN: 978-2-35158-109-4; *Ed. W. Brameshuber*

PRO 77: International Conference on 'Material Science—AdIPoC—Additions Improving Properties of Concrete—Theme 3' (2010) 459 pp., ISBN: 978-2-35158-110-0; e-ISBN: 978-2-35158-111-7; *Ed. W. Brameshuber*

PRO 78: 2nd Historic Mortars Conference and RILEM TC 203-RHM Final Workshop—HMC2010 (2010) 1416 pp., e-ISBN: 978-2-35158-112-4; *Eds. J. Válek, C. Groot, and J. J. Hughes*

PRO 79: International RILEM Conference on Advances in Construction Materials Through Science and Engineering (2011) 213 pp., e-ISBN: 978-2-35158-117-9; *Eds. Christopher Leung and K.T. Wan*

PRO 80: 2nd International RILEM Conference on Concrete Spalling due to Fire Exposure (2011) 453 pp., ISBN: 978-2-35158-118-6, e-ISBN: 978-2-35158-119-3; *Eds. E.A.B. Koenders and F. Dehn*

PRO 81: 2nd International RILEM Conference on Strain Hardening Cementitious Composites (SHCC2-Rio) (2011) 451 pp., ISBN: 978-2-35158-120-9, e-ISBN: 978-2-35158-121-6; *Eds. R.D. Toledo Filho, F.A. Silva, E.A.B. Koenders and E.M.R. Fairbairn*

PRO 82: 2nd International RILEM Conference on Progress of Recycling in the Built Environment (2011) 507 pp., e-ISBN: 978-2-35158-122-3; *Eds. V.M. John, E. Vazquez, S.C. Angulo and C. Ulsen*

PRO 83: 2nd International Conference on Microstructural-related Durability of Cementitious Composites (2012) 250 pp., ISBN: 978-2-35158-129-2; e-ISBN: 978-2-35158-123-0; *Eds. G. Ye, K. van Breugel, W. Sun and C. Miao*

PRO 85: RILEM-JCI International Workshop on Crack Control of Mass Concrete and Related issues concerning Early-Age of Concrete Structures—ConCrack 3—Control of Cracking in Concrete Structures 3 (2012) 237 pp., ISBN: 978-2-35158-125-4; e-ISBN: 978-2-35158-126-1; *Eds. F. Toutlemonde and J.-M. Torrenti*

PRO 86: International Symposium on Life Cycle Assessment and Construction (2012) 414 pp., ISBN: 978-2-35158-127-8, e-ISBN: 978-2-35158-128-5; *Eds. A. Ventura and C. de la Roche*

PRO 87: UHPFRC 2013—RILEM-fib-AFGC International Symposium on Ultra-High Performance Fibre-Reinforced Concrete (2013), ISBN: 978-2-35158-130-8, e-ISBN: 978-2-35158-131-5; *Eds. F. Toutlemonde*

PRO 88: 8th RILEM International Symposium on Fibre Reinforced Concrete (2012) 344 pp., ISBN: 978-2-35158-132-2, e-ISBN: 978-2-35158-133-9; *Eds. Joaquim A.O. Barros*

PRO 89: RILEM International workshop on performance-based specification and control of concrete durability (2014) 678 pp, ISBN: 978-2-35158-135-3, e-ISBN: 978-2-35158-136-0; *Eds. D. Bjegović, H. Beushausen and M. Serdar*

PRO 90: 7th RILEM International Conference on Self-Compacting Concrete and of the 1st RILEM International Conference on Rheology and Processing of Construction Materials (2013) 396 pp, ISBN: 978-2-35158-137-7, e-ISBN: 978-2-35158-138-4, *Eds. Nicolas Roussel and Hela Bessaies-Bey*

PRO 91 draft: CONMOD 2014—RILEM International Symposium on Concrete Modelling (2014), ISBN: 978-2-35158-139-1; e-ISBN: 978-2-35158-140-7

PRO 92: CAM 2014—2nd International Conference on advances in chemically-activated materials (2014) 392 pp., ISBN: 978-2-35158-141-4; e-ISBN: 978-2-35158-142-1, *Eds. Caijun Shi and Xiadong Shen*

PRO 93: SCC 2014—3rd International Symposium on Design, Performance and Use of Self-Consolidating Concrete (2014) 438 pp., ISBN: 978-2-35158-143-8; e-ISBN: 978-2-35158-144-5, *Eds. Caijun Shi, Zhihua Ou, Kamal H. Khayat*

PRO 94 (online version): HPFRCC-7—7th RILEM conference on High performance fiber reinforced cement composites, e-ISBN: 978-2-35158-146-9, *Eds. H. W. Reinhardt, G.J. Parra-Montesinos, H. Garrecht*

PRO 95: International RILEM Conference on Application of superabsorbent polymers and other new admixtures in concrete construction, ISBN: 978-2-35158-147-6; e-ISBN: 978-2-35158-148-3, *Eds. Viktor Mechtcherine, Christof Schroefl*

PRO 96 (online version): XIII DBMC: XIII International Conference on Durability of Building Materials and Components, e-ISBN: 978-2-35158-149-0, *Eds. M. Quattrone, V.M. John*

PRO 97: SHCC3—3rd International RILEM Conference on Strain Hardening Cementitious Composites, ISBN: 978-2-35158-150-6; e-ISBN: 978-2-35158-151-3, *Eds. E. Schlangen, M.G. Sierra Beltran, M. Lukovic, G. Ye*

PRO 98: FERRO-11—11th International Symposium on Ferrocement and 3rd ICTRC—International Conference on Textile Reinforced Concrete, ISBN: 978-2-35158-152-0; e-ISBN: 978-2-35158-153-7, *Ed. W. Brameshuber*

PRO 99 (online version): ICBBM 2015—1st International Conference on Bio-Based Building Materials, e-ISBN: 978-2-35158-154-4, *Eds. S. Amziane, M. Sonebi*

PRO 100: SCC16—RILEM Self-Consolidating Concrete Conference, ISBN: 978-2-35158-156-8; e-ISBN: 978-2-35158-157-5

PRO 101 (online version): III Progress of Recycling in the Built Environment, e-ISBN: 978-2-35158-158-2, *Eds. M. Quattrone, V.M. John*

PRO 102 (online version): RILEM Conference on Microorganisms-Cementitious Materials Interactions (2016), e-ISBN: 978-2-35158-160-5; *Eds. Alexandra Bertron, Henk Jonkers, Virginie Wiktor*

PRO 103 (online version): ACESC'16—International Conference on Advances in Civil Engineering and Sustainable Construction, e-ISBN: 978-2-35158-161-2, *Eds.T.Ch. Madhavi, G. Prabhakar, Santhosh Ram, and P.M. Rameshwaran*

RILEM Reports (REP)

Report 19: Considerations for Use in Managing the Aging of Nuclear Power Plant Concrete Structures (ISBN: 2-912143-07-1); *Ed. D. J. Naus*

Report 20: Engineering and Transport Properties of the Interfacial Transition Zone in Cementitious Composites (ISBN: 2-912143-08-X); *Eds. M. G. Alexander, G. Arliguie, G. Ballivy, A. Bentur and J. Marchand*

Report 21: Durability of Building Sealants (ISBN: 2-912143-12-8); *Ed. A. T. Wolf*

Report 22: Sustainable Raw Materials—Construction and Demolition Waste (ISBN: 2-912143-17-9); *Eds. C. F. Hendriks and H. S. Pietersen*

Report 23: Self-Compacting Concrete state-of-the-art report (ISBN: 2-912143-23-3); *Eds. Å. Skarendahl and Ö. Petersson*

Report 24: Workability and Rheology of Fresh Concrete: Compendium of Tests (ISBN: 2-912143-32-2); *Eds. P. J. M. Bartos, M. Sonebi and A. K. Tamimi*

Report 25: Early Age Cracking in Cementitious Systems (ISBN: 2-912143-33-0); *Ed. A. Bentur*

Report 26: Towards Sustainable Roofing (Joint Committee CIB/RILEM) (CD 07) (e-ISBN 978-2-912143-65-5); *Eds. Thomas W. Hutchinson and Keith Roberts*

Report 27: Condition Assessment of Roofs (Joint Committee CIB/RILEM) (CD 08) (e-ISBN 978-2-912143-66-2); *Ed. CIB W 83/RILEM TC166-RMS*

Report 28: Final report of RILEM TC 167-COM 'Characterisation of Old Mortars with Respect to Their Repair (ISBN: 978-2-912143-56-3); *Eds. C. Groot, G. Ashall and J. Hughes*

Report 29: Pavement Performance Prediction and Evaluation (PPPE): Interlaboratory Tests(e-ISBN: 2-912143-68-3); *Eds. M. Partl and H. Piber*

Report 30: Final Report of RILEM TC 198-URM 'Use of Recycled Materials' (ISBN: 2-912143-82-9; e-ISBN: 2-912143-69-1); *Eds. Ch. F. Hendriks, G. M. T. Janssen and E. Vázquez*

Report 31: Final Report of RILEM TC 185-ATC 'Advanced testing of cement-based materials during setting and hardening' (ISBN: 2-912143-81-0; e-ISBN: 2-912143-70-5); *Eds. H. W. Reinhardt and C. U. Grosse*

Report 32: Probabilistic Assessment of Existing Structures. A JCSS publication (ISBN 2-912143-24-1); *Ed. D. Diamantidis*

Report 33: State-of-the-Art Report of RILEM Technical Committee TC 184-IFE 'Industrial Floors' (ISBN 2-35158-006-0); *Ed. P. Seidler*

Report 34: Report of RILEM Technical Committee TC 147-FMB 'Fracture mechanics applications to anchorage and bond' Tension of Reinforced Concrete Prisms—Round Robin Analysis and Tests on Bond (e-ISBN 2-912143-91-8); *Eds. L. Elfgren and K. Noghabai*

Report 35: Final Report of RILEM Technical Committee TC 188-CSC 'Casting of Self Compacting Concrete' (ISBN 2-35158-001-X; e-ISBN: 2-912143-98-5); *Eds. Å. Skarendahl and P. Billberg*

Report 36: State-of-the-Art Report of RILEM Technical Committee TC 201-TRC 'Textile Reinforced Concrete' (ISBN 2-912143-99-3); *Ed. W. Brameshuber*

Report 37: State-of-the-Art Report of RILEM Technical Committee TC 192-ECM 'Environment-conscious construction materials and systems' (ISBN: 978-2-35158-053-0); *Eds. N. Kashino, D. Van Gemert and K. Imamoto*

Report 38: State-of-the-Art Report of RILEM Technical Committee TC 205-DSC 'Durability of Self-Compacting Concrete' (ISBN: 978-2-35158-048-6); *Eds. G. De Schutter and K. Audenaert*

Report 39: Final Report of RILEM Technical Committee TC 187-SOC 'Experimental determination of the stress-crack opening curve for concrete in tension' (ISBN 978-2-35158-049-3); *Ed. J. Planas*

Report 40: State-of-the-Art Report of RILEM Technical Committee TC 189-NEC 'Non-Destructive Evaluation of the Penetrability and Thickness of the Concrete Cover' (ISBN 978-2-35158-054-7); *Eds. R. Torrent and L. Fernández Luco*

Report 41: State-of-the-Art Report of RILEM Technical Committee TC 196-ICC 'Internal Curing of Concrete' (ISBN 978-2-35158-009-7); *Eds. K. Kovler and O. M. Jensen*

Report 42: 'Acoustic Emission and Related Non-destructive Evaluation Techniques for Crack Detection and Damage Evaluation in Concrete'—Final Report of RILEM Technical Committee 212-ACD (e-ISBN: 978-2-35158-100-1); *Ed. M. Ohtsu*

This report is the final report of the RILEM Technical committee 226-CNM (Concrete for Nuclear Management) which was composed of:

V. L'Hostis (CEA, France)
R. Gens (ONDRAF, Belgium), coordinator
C. Andrade (IETcc, Spain)
M. Castellote (IETcc, Spain)
C. Gallé (CEA, France)
D. Kosson (Vanderbilt University, USA)
J. Mok (CNSC, Canada)
K. Philipose (AECL, Canada)
N. Smart (Amec Foster Wheeler, UK)
P. Zuloaga (IETcc, Spain)
K. Brown (Vanderbilt University, USA)
D. Féron (CEA, France)
F. Glasser (University of Aberdeen, UK)
B. Kursten (SCK-CEN, Belgium)
J. Philip (NRC, USA)
F. Sanchez (Vanderbilt University, USA)
S. Williams (RWM, UK)

Disclaimer

The above individuals have contributed to this report through a series of formal meetings. They have reviewed the contents of the report and their review comments have been taken into account in this version of the report. Their contributions represent their own views as experts, not necessarily the view of their organizations to which they are affiliated.

Preface

Nuclear power plants and many of the facilities and structures used for the processing, the temporary storage and the disposal of radioactive waste materials generated by the fuel cycle use concrete in their construction. These structures are required to function safely and reliably in challenging and varying environments for periods of time that can potentially range up to thousands of years or even longer. During their operational life, these structures will in all likelihood be subjected to a number of environmental stresses or ageing factors that may adversely affect their performance and result in shortened service lives. The use of reinforced concrete in modern applications dates back around 100 years. Additional information is therefore required in several areas to help provide the continued assurance that these structures will continue to meet their operational requirements throughout their service life. The long-term service life and performance of concrete may be predicted with some confidence but cannot be fully validated. Analyses of archaeological analogues such as Roman concretes are a good example to build confidence in the stability of hydrated phases that can be expected in thousands of years. We can predict evolution and behaviour in the long-term but cannot validate these predictions beyond a few hundred years. It has to be acknowledged that a lot of work has already been done in this regard but key issues still have to be addressed or require additional understanding in order to build confidence.

Previous work made in RILEM by other TCs on the subject of long-term performance of reinforced concrete structures, but also in NEA/OECD related to the use of concrete structures in nuclear power plant fuel cycle facilities, pointed out several technical areas where additional knowledge was needed. The working programme of TC 226 CNM followed the TC 202 RWD managed by Dan Naus. The definition of the working programme also made use of the outcomes of the CSNI/RILEM 2004 workshop on Use and Performance of Concrete in NPP Fuel Cycle Facilities and of the NUCPERF 2006 workshop on Corrosion and Long-Term Performance of Reinforced Concrete in Nuclear Power Plants and Waste Facilities. TC 226 CNM involves interdisciplinary fields that are linked with the "core business" of RILEM and utilizes developments of several existing

RILEM committees. The activities of the Technical Committee cover the main following areas:

- Functional and performance requirements for concrete structures in the context of nuclear facilities;
- Degradation processes and their effects, particularly where the degradation phenomena can operate over extended periods of time, or where synergistic effects are present (coupling);
- Phenomenological modelling (linked to the previous point) dedicated to the long-term behaviour is problematic;
- Field experiences (collection of data during decommissioning, archaeological analogues, etc.);
- Tests approaches, instrumentation and monitoring methods dedicated to performance assessments;
- Service life models, development and validation that take into account reliability methods and updating as additional data become available (Bayesian);
- Ageing management of NPP, repair techniques to extend the performance period;
- Codes and standards specific to radioactive and hazardous waste facilities.

The main scope of the TC is the subject of R&D in several countries where nuclear industry is implemented (USA, Canada, Japan and many countries in Europe). Due to the recognized importance of the subject, it seems necessary to provide a review on existing R&D programmes on long-term prediction for nuclear applications, as well as to identify the gaps in the existing knowledge in order to propose subjects that need further research.

The Committee for Safety of Nuclear Installations (CSNI) of the OECD/NEA and the European Federation of Corrosion (EFC, Working Party 4, Nuclear Corrosion) also contributed to the activities of the TC. There was also interaction/collaboration with other projects dealing with similar topics: the Workshop on Cementitious Materials Used for Radioactive Waste Treatment, Containment, ER and D&D held in Aiken South Carolina, USA in December 2006, the International Workshop on Sulphur-Assisted Corrosion in Nuclear Waste Disposal Systems held in Brussels in 2008, and the 4th International Workshop on Long-Term Prediction of Corrosion Damage in Nuclear Waste Systems held in Bruges in 2010.

Three workshops were organized by the TC and the edited proceedings produced: NUCPERF 2009 & 2012 and AMP 2010. At the end of each workshop a summary session was conducted providing an overview of the main outcomes, questions and answers and recommendations arising during the workshops. Key input was provided by researchers from countries that are main contributors in the R&D, design, construction, operation and regulation of waste nuclear reinforced concrete facilities.

The main outcomes of RILEM TC-226-CNM are summarized in the present report.

Gif-sur-Yvette, France — Valérie L'Hostis
Brussels, Belgium — Robert Gens

Contents

Chapter 1
Some Definitions and Terminology

Robert Gens and Pablo Zuloaga

Abstract The definitions are based on standard reference documents published by Eurocode, ACI, ASTM, CEB-FIP, NEA and IAEA. Their validity and limitations are briefly discussed for major cases (NPP, interim and disposal facilities) covered by the RILEMTC 226-CNM.

Keywords Service life · Long-term behaviour · Ageing management · Durability · Safety

Abbreviations

ACI	American Concrete Institute
CEB	Euro-International Committee for Concrete
FIP	International Federation for Prestressing
FIB	International Federation for Structural Concrete
IAEA	International Atomic Energy Agency
NEA	Nuclear Energy Agency
ASTM	American Society of the International Association for Testing and Materials
NPP	Nuclear Power Plant
ICRP	International Commission on Radiological Protection
ILW	Intermediate Level Waste
LILW	Low and Intermediate Level Waste
LLW	Low Level Waste

R. Gens (✉)
ONDRAF-NIRAS, Brussels, Belgium
e-mail: r.gens@nirond.be

P. Zuloaga
ENRESA, Madrid, Spain

V. L'Hostis and R. Gens (eds.), *Performance Assessment of Concrete Structures and Engineered Barriers for Nuclear Applications*,
RILEM State-of-the-Art Reports 21, DOI 10.1007/978-94-024-0904-8_1

1.1 Service Life

Service life is defined in general terms as the time period in which the structure fulfils the design requirements. It is verified by mathematical expressions named "limit state functions" which separate the complying state from the undesired one. Then, the key aspect is to define the requirements and to express the several ones by simple limit state functions, which could be verified by modelling, by testing or by direct observations.

Limit states are states of the structure defining unfitness for use. Limit states are the states beyond which the structure no longer fulfils the relevant design criteria. Two different types of limit states are classically considered, namely Ultimate Limit State and Serviceability (functional) Limit State (CEB-FIP Model Code 1990).

Attainment of the bearing capacity of a structural part or of the structure as a whole is classified as an ultimate state. The serviceability limits states associated with the general requirements refer to (CEB-FIP Model Code 1990):

- Limited local structural damage such as excessive cracking or excessive compressive stresses, producing irreversible strains and microcracks
- Deformations which produce unacceptable damage in non-structural elements or excessively affect the use or appearance of structural or non-structural elements

The following definitions are reported:

- Eurocode (Eurocode0 2005):
  - Service life is the period of time during which the structure maintains the design requirements on:

    Safety
    Functionality and
    Aesthetics

  without unexpected costs of maintenance.

- Design working life: is the period for which a structure or part of it is to be used for its intended purpose with anticipated maintenance but without major repair being necessary
- Durability: the durability of a structure or part of it in its environment is such that it remains fit for use during the design working life given appropriate maintenance
- ACI (2000):
  - Durability is the capability of maintaining the serviceability of a product, component, assembly, or construction over a specified time. Serviceability is viewed as the capacity of the above to perform the function(s) for which they are designed and constructed
  - Service life (of building component or material) is the period of time after installation (or in the case of concrete, placement) during which all the

properties exceed the minimum acceptable values when routinely maintained. Note, however, that it may not be possible to achieve the expectation of this definition—that maintenance is possible. Three types of service life have been defined (Sommerville 1986 cited in ACI 2000):

Technical service life is the time in service until a defined unacceptable state is reached, such as spalling of concrete, safety level below acceptable, or failure of elements
Functional service life is the time in service until the structure no longer fulfils the functional requirements or becomes obsolete due to change in functional requirements, such as the needs for increased clearance, higher axle and wheel loads, or road widening. It should be mentioned that degradation appearance does not mean loss of functionality
Economic service life is the time in service until replacement of the structure (or part of it) is economically more advantageous than keeping it in service

The terms "durability" and "service life" are often erroneously interchanged. The distinction between the two terms is evident when their definitions, as given in ASTM E 632, are compared (ACI 2000).

- IAEA glossary (IAEA 2007):
  - Design life: the period of time during which a facility or component is expected to perform according to the technical specifications to which it was produced
  - Operating life/lifetime: the period during which an authorised facility is used for its intended purpose, until decommissioning or closure. The synonyms operating period and operational period are also used. The period during which a spent fuel or a radioactive waste management facility is used for its intended purpose. In the case of a disposal facility, the period begins when spent fuel or radioactive waste is first emplaced in the facility and ends upon closure of the facility
  - Qualified life: period for which a structure, system or component has been demonstrated, through testing, analysis or experience, to be capable of functioning within acceptance criteria during specific operating conditions while retaining the ability to perform its safety functions in a design basis accident or earthquake
  - Service life: the period from initial operation to final withdrawal from service of a structure, system or component
  - Life management (or lifetime management) in which due recognition is given to the fact that at all stages in the lifetime there may be effects that need to be taken into consideration. An example is the approach to products, processes and services in which it is recognised that at all stages in the lifetime of a product (extraction and processing of raw materials, manufacturing, transport and distribution, use and reuse, and recycling and waste management) there are environmental and economic impacts. The term 'life

cycle' (as opposed to lifetime) implies that the life is genuinely cyclical (as in the case of recycling or reprocessing)
- Ageing management: see Sect. 1.3

The available models for assessing the service life can be classified in different manners depending on whether the mathematics or the number and type of parameters are considered. They may be:

- Analytical/numerical
- Empirical/or test-based (Durability Indicators)
- Probabilistic
- Multi-scale (micro to macro)

Given the definitions mentioned above, to define the service life of a nuclear installation, one needs to define the corresponding requirements. This raises the question of what limit states have to be accounted for. Each kind of relevant structural component involved in LILW facilities will have different requirements depending on their purpose and length of operation.

In most cases the basic criteria adopted for surface disposal of LILW include a design objective of a few hundred years' service life for the engineered barriers and low maintenance requirements. As isolation of radionuclides is the main function of these barriers, there is no clear definition of the limit states: strictly, they should be interpreted in terms of dose restriction (e.g. dose constraint to the population no more than about 0.3 mSv/y) (ICRP 2007), although, for practical reasons, service life is normally addressed in more conventional structure-related durability criteria. The durability required for such extended periods makes it necessary to discuss confidence building rather than demonstration. Confidence may be supported not only by modelling but by proposing robust solutions, together with a good knowledge of the processes involved. These arguments are presented to the stakeholders through a safety case, containing a Performance Assessment with simplified models using envelope assumptions. The goal of the ageing management program, in the operational phase of the facility, is to better understand all the potential processes and to confirm that the hypotheses adopted in the Performance Assessment are really envelope conditions, thus improving the long-term safety case and the knowledge of the existing margins (Zuloaga et al. 2010).

Service life cannot therefore be defined in a single way which is valid for all cases. A distinction has to be made between short-term facilities (e.g. NPP, interim storage), which have a well-defined required lifetime, and long-term disposal facilities, which will remain indefinitely. In the latter case, the notion of service life is far from being limited to the physical and/or mechanical performance of the disposal system. The role of the chemical barrier is at least as important as structural considerations. For disposal facilities, the radiological issue (i.e. the acceptable dose threshold) is the determining factor for long-term safety. Service life depends on the field of application—it is the period during which the structure fulfils the design requirements—which are very different for a NPP or a disposal facility.

1.2 Long-Term Behaviour

Long-term operation (LTO) for NPPs is operation beyond the established timeframe originally stipulated in the license terms, design limits, standards or regulations for the plant. LTO needs to be justified by a safety assessment that considers life-limiting processes and features for structures, systems and components. Proper and safe LTO is based on the experience and practices of various countries in areas such as plant license renewal, life extension, continued operation and life management. Other activities, including periodic safety review, ageing management and plant modification, are also relevant to LTO (IAEA 2008).

The following definitions related to assessment are given in the IAEA glossary (IAEA 2007):

- Assessment: the process, and the result, of analysing systematically and evaluating the hazards associated with sources and practices, and associated protection and safety measures. Assessment is often aimed at quantifying performance measures for comparison with criteria. In IAEA publications, assessment should be distinguished from analysis
- Assessment is aimed at providing information that forms the basis of a decision on whether or not something is satisfactory. Various kinds of analysis may be used as tools in doing this. Hence an assessment may include a number of analyses
- Performance assessment: assessment of the performance of a system or subsystem and its implications for protection and safety at an authorised facility. This differs from safety assessment in that it can be applied to parts of an authorised facility (and its environment), and does not necessarily require the assessment of radiological impacts
- Risk assessment: Assessment of the radiological risks associated with normal operation and possible accidents involving a source or practice. This will normally include consequence assessment, together with some assessment of the probability of those consequences arising
- Safety assessment: Assessment of all aspects of a practice that are relevant to protection and safety; for an authorised facility, this includes siting, design and operation of the facility. This will normally include risk assessment. Analysis to predict the performance of an overall system and its impact, where the performance measure is the radiological impact or some other global measure of the impact on safety. The systematic process that is carried out throughout the design process to ensure that all the relevant safety requirements are met by the proposed (or actual) design. Safety assessment includes, but is not limited to, the formal safety analysis

For disposal systems, because of the timescale concerned (from a few hundred years for surface disposal up to hundreds of thousands years for deep geological disposal), the evaluation of the long-term behaviour of the repository system requires the development of an assessment basis. The assessment basis is the collection of information and analysis tools supporting the safety assessment. This includes an overall description of the disposal system, the scientific and technical data, understanding relevant to the assessment of system safety and feasibility, assessment methods, models, computer codes and databases for analysing system performance. Relevant scientific and technical data and understanding can come from well-established physical and chemical principles, rock characterisation studies, desk studies, laboratory studies, full-scale demonstrations, existing industrial applications, or experience within and outside the nuclear industry (NEA 2013).

1.3 Ageing Management

The IAEA gives the following general definition for ageing management: engineering, operations and maintenance actions to control within acceptable limits the ageing degradation of structures, systems and components. Examples of engineering actions include design, qualification and failure analysis. Examples of operations actions include surveillance, carrying out operating procedures within specified limits and performing environmental measurements.

Life management (or lifetime management) is the integration of ageing management with economic planning: (1) to optimise the operation, maintenance and service life of structures, systems and components; (2) to maintain an acceptable level of performance and safety; and (3) to maximise the return on investment over the service life of the facility.

An Ageing Management Program (AMP) is a set of policies, processes, procedures, arrangements, and activities for managing the ageing of the systems, structures and components (SSCs) for an NPP. AMP ensures an adequate framework for coordinating programs and activities relating to the understanding, control, monitoring and mitigation of ageing of a plant component. AMP does not replace existing programs; however, on the basis of evaluation, it modifies them to achieve a systematic and integrated program for effective AM. A key issue is also the qualification and training of the personnel (see Chap. 4 "Ageing Management of Concrete Structures of NPPs").

References

ACI: Service-life prediction—state-of-the-art report, ACI 365. 1R-00 (2000)

CEB-FIP Model Code 1990: Comité Euro-International du Béton, Published by Thomas Telford Services Ltd first published 1993, Reprinted 1998, ISBN: 0-7277-1696-4

Eurocode0: Basis of structural design, EN 1990:2002+A1 (2005)

IAEA: IAEA Safety Glossary, Terminology Used in Nuclear Safety and Radiation Protection. 2007 Edition Safety Glossary (2007)

IAEA: SALTO Guidelines, Guidelines for Peer Review of Long-Term Operation and ageing Management of Nuclear Power Plants. Services Series 17 (2008)

ICRP: The 2007 Recommendations of the International Commission on Radiological Protection. Ann. ICRP **37**, 2–4 (2007)

NEA: The Nature and Purpose of the Post-Closure Safety Cases for Geological Repositories. Safety Case Brochure 2012. NEA/RWM/R(2013)1, 25 Mar 2013 (2013)

Zuloaga, P., Ordonez, M., Andrade, C., Castellote, M.: Ageing management program for the Spanish LLW disposal and spent fuel and high-level waste centralised storage facilities. In: L'Hostis, V, Philipose, K, Gens, R, Gallé, C (eds.) AMP 2010—International Workshop on Ageing Management of Nuclear Power Plants and Waste Disposal Structures (EFC Event 334), Toronto, Ontario, Canada, 7–10 Nov 2010. ISBN:978-2-7598-0636-2, vol. 12 (2011), EPJ Web of Conferences 12,01003(2011), doi:10.1051/epjconf/20111201003, published by EDP Sciences (2010)

Chapter 2
Materials and Properties

Kevin Brown, Christophe Gallé, David Kosson, Florence Sanchez and Jean Michel Torrenti

Abstract Various cementitious materials are used in nuclear applications covering a broad range of environmental conditions and time scales. The characterisation of cementitious materials and the determination of their properties are a key issue for assessing the evolution of nuclear infrastructures. This section gives in first place an overview of blended cements, low pH-cementitious materials and fiber reinforced and advanced cement-based materials. Degradation of cement-based materials in their environments is to a significant extent controlled by their transport properties. For this purpose, reactive transport models are being developed integrating multiple coupled phenomena. The scientific assessment basis needed for the development of these models is discussed. In particular, leaching, carbonation and oxidation of cementitious materials are dealt with. Besides chemical processes, desiccation and pre-cracking play also an important role in determining the transport properties of cementitious materials. Scaling to realistically-sized engineered systems under field conditions is a challenging issue.

Keywords Transport properties · Hydration · Carbonation · Leaching · Desiccation · Pre-cracking · Coupled phenomena

2.1 Introduction

Cementitious materials are used in reinforced concrete for structural components, as well as in grouts and wasteforms in nuclear applications. The context for use is also varied including nuclear power facilities, used fuel interim storage and final

---

K. Brown · D. Kosson (✉) · F. Sanchez
Vanderbilt University, Nashville, USA
e-mail: david.kosson@vanderbilt.edu

C. Gallé
CEA, Saclay, France

J.M. Torrenti
IFSTTAR, Paris, Champs Sur Marne, France

V. L'Hostis and R. Gens (eds.), *Performance Assessment of Concrete Structures and Engineered Barriers for Nuclear Applications*,
RILEM State-of-the-Art Reports 21, DOI 10.1007/978-94-024-0904-8_2

repository disposition, nuclear-chemical processing, facility closure and waste management. Extensive use of blended cements is anticipated in the various components of future nuclear facilities to achieve improved performance characteristics such as higher strength, durability, and radionuclide retention. This initiative is also guided by the global need to optimise the use of raw materials in the near future and to limit the CO_2 foot-print of industrial activities. It is notable that Roman cement is still performing well in many structures after a couple thousand years. As such, large quantities of coal fly ash, ground granulated blast furnace slag (GBFS), and/or silica fume may be used as supplemental cementitious materials along with reduced quantities of Portland cement. Current performance assessment tools, however, have been developed for systems composed mostly of Portland cement. There is, therefore, a critical need for tools that can predict the properties and performance of systems composed of a broad range of cementitious materials. This is a key issue when integrating the long-term aspect for nuclear infrastructure.

2.2 Blended Cements and Hydrated Phase Prediction

Prediction of properties and system performance at longer ages is handicapped by the lack of data with which to implement and to validate the predictive modelling. In particular, the decline in pH over time is uncertain and the mineralogical evolution of the system is uncertain. It is, therefore, important to overcome these knowledge deficiencies. An important advance in knowledge would be the ability to define the hydrated phases produced during the hydration reactions of blended cements, and their associated physical, chemical, thermodynamic and kinetic characteristics and along with the pore solution composition as they play a critical role in the mechanical strength, buffering capacity and radionuclide retention properties, and, thus, impact the overall system performance.

Recently, Snyder et al. (2009) proposed a phase development model in which standardised materials characterisation techniques are used in conjunction with the thermodynamic hydration model developed by Lothenbach et al. (2008) to validate the hydrate phase prediction. Snyder et al. (2009) found X-ray powder diffraction (XRD) with Rietveld analysis and thermo-gravimetric analysis (TGA) to be two very useful techniques in identifying and quantifying hydrated cementitious phases for a range of cementitious binder mixtures anticipated for nuclear applications. Arnold et al. (2012) evaluated the use of Portland cement hydrated phase definitions for predicting solid-liquid partitioning of major constituents from a fly ash, slag, and Portland cement mortar after 30 months of curing and also carried out quantitative scanning electron microscopy—energy dispersive x-ray (SEM-EDX) characterisation of the material. Results indicated that earlier phase definitions provided a useful starting point but modelling of the observed behaviour was challenging for silica at pH above 12 and for aluminium at pH less than 12. Only fractional reaction of GBFS and fly ash was observed experimentally and phyllosilicates, likely introduced with the sand fraction, appeared to play an important role in partitioning of some constituents. Further research is needed to quantify the reaction rate of fly ash and slag and their respective hydration products. Independent validation of the resulting phase quantification is needed by a combination of approaches such as XRD, TGA and SEM-EDS.

Hidalgo López et al. (2008) also demonstrated the usefulness of SEM and infrared spectroscopy for phase definition in blended cements. Nuclear magnetic resonance (NMR) spectroscopy has been shown as a useful tool in evaluating silicate polymerisation (Cong and Kirkpatrick 1996; Colombet et al. 1998; Ramachandran and Beaudoin 2001). While the proposed approaches show promise, important remaining issues include the determination of phases that should be included in the predictive model and the confidence that should be placed in the selection of these phases. Furthermore, the inclusion of GBFS in the material formulation imposes chemical reducing properties within the cement matrix (often used to reduce mobility of technetium), which may slowly oxidize with the ingress of atmospheric oxygen. Under these conditions, the changes in speciation of several important constituents within the cement matrix, such as sulphur and several radionuclides, remains uncertain.

2.3 Characterisation of Low pH-Cementitious Materials

Underground repositories for long-term disposal of high-level radioactive waste are based on the concept of a multiple-barrier system to ensure waste isolation for tens or hundreds of thousands of years. Several designs include a clay-based engineered barrier such as bentonite in contact with large amounts of cementitious materials. Pore waters originating from Portland cement-based materials have a high alkalinity ($pH > 13$) and are able to react with and modify the mineralogy and properties of the bentonite barrier (Fernández et al. 2009). The alteration of bentonite is characterised by montmorillonite dissolution at the early stage and by the substitution of zeolites presumably by cement phases such as calcium silicate hydrates (C–S–H) on the long-term (Fernández et al. 2009). The use of low pH cementitious materials ($pH < 11$) is an accepted method to ensure bentonite stability over long periods of time. Calcium aluminate cements (CAC) are thus an interesting alternative to Portland cement-based materials to prevent reaction with bentonite barriers. However, CAC suffer from hexagonal calcium aluminate hydrate phase conversion, generally resulting in an increase in porosity and loss of strength.

Addition of high silica content admixtures such as silica fume has shown to be effective in reducing the hydrate conversion process, increasing the stability of the microstructure, and slightly decreasing further the pore water pH of the cementitious material (Hidalgo López et al. 2008; Garcia Calvo et al. 2009). The microstructure of the new binder is characterised by the development of new phases such as strätlingite (C_2ASH_8) and siliceous hydrogarnet. While these new binders show promise for use in the underground facility of a nuclear waste repository, little is known on their properties and impact on long-term performance.

2.4 Fiber Reinforced and Advanced Cement-Based Materials

Randomly oriented nano/microfiber (steel, carbon, polymeric, and glass) reinforced cement-based materials are an important class of composite materials that can be tailored for specific applications and can open the door for new applications in nuclear waste containments. The use of these fibers improves the post cracking load bearing capability of cement-based materials by controlling the growth of cracks, limiting the crack width, and improving the material energy absorption performance, resulting typically in an increase in strength, toughness, impact resistance, fatigue strength, and durability as well as a reduction in plastic shrinkage cracking (Banthia and Sheng 1996; Katz 1996; Lange et al. 1996; Banthia and Nandakumar 2003; Shah et al. 2004). Fiber reinforcements provide additional, unique properties, including low electrical resistivity, electromagnetic field shielding, self-sensing capabilities (carbon and steel) (Chung 2000, 2002; Reza et al. 2003), high ductility and self-control of cracks (polymer) (Li 2003), making them attractive for applications where the long-term performance and advanced monitoring of the structure is critical. Whereas, a large number of studies have been conducted to examine the direct structural, mechanical, and electrical properties provided by the fibers (Wang et al. 1987; Katz and Bentur 1994; Toutanji et al. 1994; Fu and Chung 1995; Banthia and Sheng 1996; Lange et al. 1996; Torrents et al. 2000; Nelson et al. 2002), the long-term chemical and structural stability of these materials in response to severe conditions such as thermal and radiation environments and environmental weathering has received little attention.

Among the various properties that influence the composite material, the interfacial bond between cement paste and fiber is the most important. Stress (and thus load) transfer between the cement-based material and the reinforcement takes place through the interface and interfacial zone. It is the efficacy of adhesion that determines the load that can be distributed to the fibers, ultimately determining the maximum load-bearing capacity of the composite structure. Over its lifetime, the material is subjected to a multitude of physical, chemical, and mechanical degradation processes, causing internal chemical changes and stresses that directly affect the interfacial bond between the reinforcing fibers and cement (Sanchez and Borwankar 2010). Pull-out of fibers during cracking can result in exposed fibers in cracks during degradation. Prediction of the long-term performance of fiber reinforced cement-based composites requires, therefore, understanding how the fiber-cement interface changes as a function of material weathering and resultant chemistry.

Decalcification is of critical concern in structures used for radioactive waste disposal and is closely associated with various types of concrete deterioration. A study of the effect of decalcification on the fiber-cement interface and in turn the mechanical properties of carbon microfiber reinforced cement-based composites has been performed by Sanchez et al. (2009), Sanchez and Borwankar (2010). Results showed a strong coupling between calcium leaching at the fiber-cement interface and the mechanical degradation of the composites. Prior to decalcification, the fiber cement interface in Portland cement pastes was characterised by the presence of a thin layer rich in calcium hydroxide. During exposure to ammonium nitrate solution, preferential leaching around the fibers occurred resulting in debonding of the fibers and a greater loss of strength. However, as the composites leached, the fibers increasingly contributed to an increased ductility. Decalcification changed the failure mode from brittle cracking to slow ductile load dissipation, which was more pronounced for the Portland cement pastes with the fibers. Addition of silica fume to the mix slowed the degradation process, stabilised the fiber-cement interface during decalcification, and reduced the loss of strength. Research is needed, however, to investigate the effect on the fiber integrity of high shear rates to homogenise samples with silica fume.

The use of nano-sized (e.g., nano-silica, nano-calcium carbonate, and nano-iron) and nanostructured (e.g., nanoclay) particles and also nanotubes and nanofibers have been shown to enhance the performance of cement-based materials, providing higher compressive and flexural strengths, improved hydration and internal curing characteristics, early cracking resistance, and higher durability compared to conventional cement-based materials (Raki et al. 2010; Sanchez and Sobolev 2010; Kawashima et al. 2013; Pacheco-Torgal et al. 2013). These innovative concretes are thus promising for the containment of radioactive wastes, and research is needed to evaluate their long-term performance under environmental conditions relevant to nuclear waste storage. Recent efforts in this area include a study of the effect of decalcification on the chemo-mechanical behaviour of cement pastes containing carbon nanofibers (CNFs) (Brown et al. 2012b). Results showed a strong correlation between the CNF dispersion state, the microstructural evolution of the cement paste during decalcification, and the material mechanical properties.

2.5 Transport Properties

Description of the degradation of cement-based materials subjected to exposure to air, water, and saline solutions requires consideration of the rate-limiting mass transfer processes that control the distribution of reactants within the material, chemical speciation, equilibrium, and reaction kinetics between the different phases (henceforth referred to as reactive transport modelling). An initial step in developing a reactive transport model for performance of a cement-based material is the definition of a conceptual model of the material and its associated system, including important transport mechanisms and pathways (e.g., liquid and vapour phase diffusion, percolation, capillary transport, cracking and preferential flow pathways), system chemistry (e.g., aqueous-solid and aqueous-vapour partitioning, mineral phases, local equilibrium vs. kinetic representation), and system initial and boundary conditions (e.g., water contact and infiltration, relative saturation, vapour, liquid and solid interfaces). One example for system definition and interactions is provided in Fig. 2.1. Reactive transport modelling then requires either direct

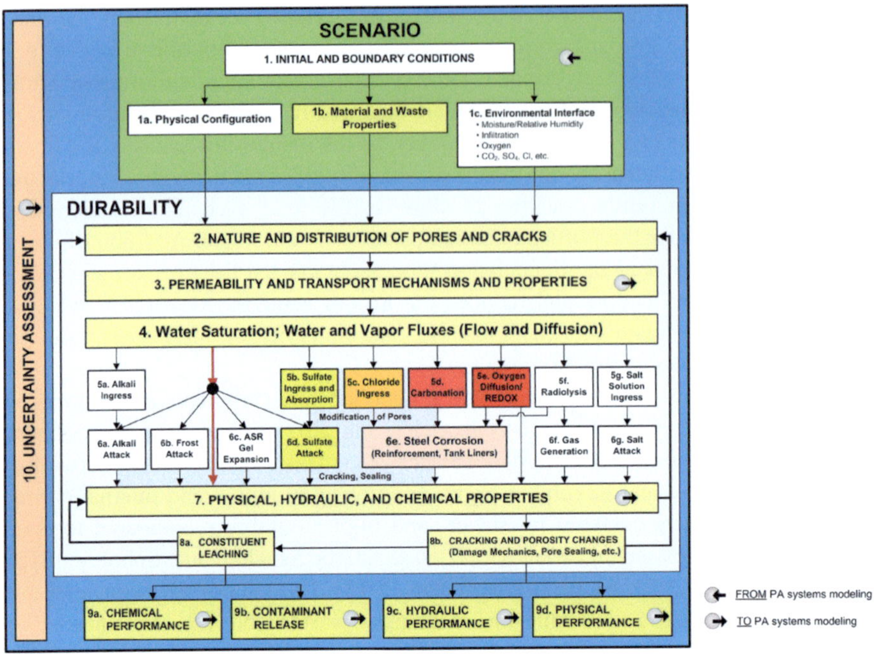

Fig. 2.1 One example of system definition and interactions for evaluation of cementitious materials performance in nuclear applications

measurement or simulation-based quantification of key reactive transport parameters that are related to the specific material composition and physical structure, as well as saturation and temperature. Extensive reviews of conceptual models, constituent transport, and degradation mechanisms for cement materials used in nuclear applications are available (CBP 2009; Pabalan et al. 2009; Weiss 2012).

Research associated with transport and reaction of constituents through or from concrete can be considered as (i) providing an improved microscale understanding of various phenomena leading to improved conceptual and mathematical models, (ii) providing needed parameter estimates using measurement or simulation based approaches (Garboczi et al. 1999; Martys and Hagedorn 2002; Arns et al. 2004; Ukrainczyk et al. 2012), and (iii) model scaling and simplification to evaluate engineering-scale applications and scenarios.

At the microscale, recent research has emphasised the relationships between microstructure, sand fraction, and effective diffusivity as a function of water to cement ratio relative the importance of the interfacial transition zone (ITZ) in understanding observed aqueous diffusion processes (Larbi et al. 2012). Results suggest that the more porous ITZ around individual aggregate particles becomes more connected at sand volume fractions greater than approximately 50 %, resulting in significantly increased aqueous phase and gas phase (under unsaturated conditions) diffusivity.

Another consideration is the potential influence of the electric double layer (EDL) that forms around the negatively charged cement hydration products to restore electroneutrality. The EDL has been reported to have a non-negligible effect on the transport of ionic species (Chatterji 1994a, b; Revil 1999; Yu and Page 1991; Castellote and Llorente 2006). It has been reported that the effect of the EDL is more significant at low water/cement ratios than high water/cement ratios (Chatterji 1994a). Elakneswaran et al. (2009, 2010) investigated the influence of the EDL properties on chloride ingress in cement-based materials and proposed an integrated thermodynamic modelling approach that incorporates phase equilibrium and surface complexation models in addition to multicomponent diffusion in charge free pores and diffuse double layer of hydrated cement pastes. C–S–H, Portlandite, ettringite, and monosulfate phases were considered for the equilibrium reactions and the surface complexation reactions were limited to C–S–H Blended cements were used in the study, including a hydrated cement paste, a slag cement paste, and a fly ash cement paste. The sorptive properties of the cement pastes were explained on the basis of the classical Freudlich theory. The diffusion through the EDL was modelled considering the linearised Poisson-Boltzmann equation for symmetric electrolytes. Results showed that the EDL properties of the hydrated cement pastes had a significant effect on the adsorption and transport of chloride through the gel pores (diameter less than 10 nm), which occupied more than 30 % of the total pore volume of the pastes. The slag cement paste had the lowest porosity and the greatest amount of gel pores and showed the greatest ability of chloride to bind both chemically and physically, resulting in the greatest influence on the ingress of chloride as compared to the other pastes.

Arnold et al. (2013) used a solution of the full non-linear Poisson-Boltzmann equation for symmetric and asymmetric electrolytes and illustrated that use of the linearised Poisson-Boltzmann equation results in approximation errors on the order of 10 % for symmetric electrolytes and up to 50 % for asymmetric electrolytes with the greatest effect of the EDL in gel pores. In summary, while it is recognised that ionic transport in cementitious materials is affected by the intercoupling of physical and chemical phenomena, detailed transport mechanisms of ionic species are still not well understood and the known effect of the EDL is generally not accounted for in reactive transport models. As a result, calculated diffusion coefficient values determined only based on Fick's law without consideration of other ionic effects, increases the uncertainties in predicting long-term aggressive substance ingress into cementitious materials.

Development of simulation-based approaches for estimating relevant effective transport parameters has been on-going (Garboczi et al. 1999; Martys and Hagedorn 2002; Arns et al. 2004; Ukrainczyk et al. 2012). Efforts in this area have been focused on developing appropriate three-dimensional digital images or virtual representations of pore structure, either from direct imaging or through rule-based hydrated microstructure assemblage formation. This latter modelling approach provides information of fundamental importance in terms of mechanistic understanding of transport parameters such as porosity, pore interconnectivity, and tortuosity; however, how these parameters change with time during exposure needs to be considered. Efforts are also needed with respect to parameter identification and their degree of certainty to ensure representation of the reality.

2.6 Coupled Phenomena and Materials Behaviour

As illustrated earlier in Fig. 2.1, understanding performance of cementitious materials in realistic engineered applications[1] requires integration of multiple phenomena, many of which are inextricably coupled, and assumed scenarios regarding the expected environment in which the materials are used. Among the materials challenges for cementitious materials that have been identified for the next decades, optimisation of material properties and durability, extension of the service life of civil engineering structures, reduction of CO_2 emissions, and sustainable use and management of resources are certainly the most important ones. Addressing these issues is essential when considering the long-term functioning and durability requirements for nuclear construction (e.g., reinforced concrete confinement building) and waste management components (e.g., waste containers, wasteforms, disposal systems). Three aspects of coupled phenomena and application are cross-cutting amongst most applications: (i) carbonation, (ii) leaching, and (iii) scale-up and integration for evaluation of realistic engineered systems.

2.7 Carbonation and Oxidation of Cementitious Materials

Carbonation of cementitious materials, from reaction with atmospheric or other sources of carbon dioxide such as biogenic, can result in a decrease in the pore-water pH of cementitious systems to pH ~ 9 which can lead to depassivation of embedded steel and thus to the onset of steel reinforcement corrosion for structural materials and increased leaching of radionuclides or other constituents. The rate of carbonation is a function of the following: (i) material alkalinity and pore structure (including potential crack pathways), (ii) water saturation, and (iii) concentration of external carbon dioxide sources. Several NUCPERF papers focused on improving the understanding of carbonation processes. Brown et al. (2012a) carried out reactive transport modelling sensitivity studies focused on the impacts of concrete formulation and exposure conditions for underground high level waste storage tanks at the U.S. Department of Energy sites. For simulation, Stora et al. (2009) performed numerical modelling of unsaturated accelerated carbonation of CEM I cement-based materials using the DUSS software. The simulations underlined the importance of

[1]cementitious materials performances adequacy with their functionality for nuclear engineered applications.

initial mineral phase composition. In particular, the study showed that the reduction of calcium-bearing hydrates due to the substitution of clinker by silica enhances the carbonation propagation, even though the initial porosity slightly decreased. Thouvenot et al. (2012) presented the results of coupled cement drying and carbonation in the context of waste packages for intermediate level waste in deep geologic disposal. Results indicated that the drying period for 11-cm thick concrete components was between 2 and 10 years, and carbonation depths on the order of 2–3 cm are expected over 100 years. Arnold et al. (2009) described the effects of carbonation on constituent retention in a cementitious wasteform and Auroy et al. (2012) described ongoing experimental studies on the impacts of the carbonation process on transport properties of the resultant cementitious matrix.

An analogous reaction is the oxidation of wasteforms through gas phase diffusion of atmospheric oxygen or introduction as dissolved oxygen in water and reaction at the water film-solid interface. Reducing admixtures, such as granulated blast furnace slag, are added to cementitious wasteforms to maintain reducing conditions as a mechanism for increased retention of some important radionuclides, such as technetium-99 (Pabalan et al. 2012). Pabalan et al. (2012) studied the rates of leaching as a consequence of oxidation for one wasteform composition; however, the rate and extent of oxidation, along constituent retention, are not generally well known. For example, it is unclear if release is a function of a rind formation around reduced particles or if oxidation is more uniform.

These studies point to the research and application needs of (i) gas phase diffusivity and reaction as a function of water saturation; (ii) improved definition of the mineral phases and associated thermodynamic solubility parameters for carbonated materials (especially for compositions with supplemental cementitious materials such as fly ash, slags, and silica fume); (iii) clearer understanding of anticipated oxygen, carbon dioxide, and relative humidity boundary conditions under anticipated exposure conditions; and (iv) the need to understand the impact of changes in mineralisation as a result of carbonation and oxidation on transport properties.

2.8 Leaching

Leaching processes and assessment are important because leaching of major material constituents can result in decalcification and decrease in pore solution pH leading to corrosion and degradation of structural or hydraulic properties. Leaching of trace species results in the release of radionuclides and other constituents of potential concern. Recently, the U.S. Environmental Protection Agency has standardised a set of four leaching test methods, developed as part of the Leach Environmental Assessment Framework (LEAF), that are suitable for evaluation of cementitious materials and wasteforms. The goal of the LEAF methods is to describe constituent leaching as a function of one or more release-controlling parameters rather than to simulate leaching under a single set of test conditions.

The LEAF leaching methods have been documented (Kosson et al. 2002; Garrabrants et al. 2010), validated (Garrabrants et al. 2012a, b), and are now adopted as EPA testing methods (USEPA 2013). Tests conducted using the LEAF approach include methods to characterise (i) liquid-solid partitioning (LSP) as a function of pH using EPA Method 1313, (ii) LSP as a function of liquid-to-solid ratio (L/S) using EPA Method 1314 or Method 1316, and (iii) mass transfer parameters as a function of leaching time using EPA Method 1315. These methods not only support U.S. regulatory applications, but also provide a basis for parameterising reactive transport models for concrete and cementitious wasteforms (Sarkar et al. 2010, 2012; Brown et al. 2012a). Currently, these methods are being used to evaluate formulations for secondary and low activity wasteforms as part of waste management at former defense sites in the U.S. (Mattigod et al. 2011; Um et al. 2011; Arnold et al. 2012).

Within the framework of the development of the LILW disposal facility of El Cabril in Spain, a service life model was developed (Zuloaga et al. 2009). A specific study dedicated to the leaching of concrete barriers of the disposal vaults taking into account the evolution of water content due to seasonal temperature changes was carried out. The experience showed that the engineered barriers behaved as assumed with previous estimation except in the case with the temperature changes. It was concluded that, the life assessment of a real engineered barrier has thus to consider all realistic scenarios and taking into account all the materials and the elements of the multi-barrier isolation system.

Another example of correlations between measures and predictive model development for long-term behaviour of concrete structures subjected to leaching was provided by a study carried out by de Larrard et al. (2009). This work was the opportunity to compare various field measurements achieved during concrete structure building operations (tunnel and bridge), with calculations performed with a finite volume method. Numerical simulation of accelerated leaching test using ammonium nitrate were carried-out with a special focus on the effect of the mesh refinement on the solid calcium concentration profile and the optimisation of computational times. The finite volume modelling method developed within this study was regarded as a very promising approach to provide probabilistic calculations taking in account the variability of calcium diffusion parameters.

Garcia Calvo et al. (2009) investigated the durability of low-pH cementitious materials, specifically their resistance to long-term aggression to groundwater from real repository conditions. Leaching tests at low overpressures ($\sim$0.5 bars) and groundwater from the Aspo site (Sweden) with high Cl^- and Mg^{2+} content were used. The use of silica fume and fly ash as high silica content admixtures was evaluated. Results showed that the material based on CAC plus silica fume provided a good resistance to groundwater aggression while that based on CAC plus fly ash showed an altered front with decalcification of the C–A–S–H phases and incorporation of magnesium ions. The materials tested had, however, relatively high porosity that could be due to the formation of dense hydrogarnet and the use of a high water to binder ratio.

Other recent work by Dauzères et al. (2010a, b) on clay-rock/cement-based material interaction in the context of geological disposal has provided interesting results through the comparison of the leaching behaviour of CEM I and low pH-pastes. Leaching tests (at the cement/clay interfaces while exerting confining pressure on the solid) were performed for a minimum of one year using synthetic mudstone water at neutral pH and 25 °C under a controlled CO_2 environment (1.3 %). Results showed that the decalcification of the low pH-material was greater than that of CEM I, showing the formation of amorphous silica and an intense capillary porosity opening. It was also observed that the carbonation of the CEM I paste was quite limited with only the formation of a superficial calcite layer. In contrast, the low pH-paste was carbonated over the entire decalcified zone. For this material, a magnesium enriched zone (M–C–S–H) behind the decalcified and carbonated domain was also observed.

The results described above show that there is still significant research needed to understand and predict the long-term behaviour of complex materials in the disposal context. Research would benefit from well described reference cases, use of standardised leaching characterisation methods to allow comparison of results in combination studies designed to elucidate specific ageing mechanisms, and the resultant impacts on constituent release and chemical, mechanical, and transport properties. Additional research is also needed that connects small-scale laboratory results with observations of field test systems over prolonged timeframes.

2.9 Desiccation and Pre-cracking in Cement Pastes and Mortars

During their lifetime, cracking in concrete structures could occur due to mechanical loading (at the service state a limited opening depending on the standards and the environmental conditions is allowed for reinforced concrete structures) or due to physical (autogenous, thermal and drying shrinkage, freeze-thawing, elevated temperatures…) or chemical phenomenon (alkali-aggregate reaction, delayed ettringite formation…) and finally lead to a decrease in durability.

At early age, in massive structures, if shrinkage is restrained or due to temperature gradients, internal stresses develop and cracks could occur. This is an important problem that could be experimentally tested for instance by means of an active restrained shrinkage ring test (Briffaut 2009). This test allows comparing the effect of different concrete mixes or different reinforcement on the cracking behaviour. In this test, all phenomenon involved (evolution of mechanical properties with hydration, thermal and autogenous shrinkages, creep, mechanical damage and their couplings) are considered. The behaviour of massive concrete structures is actually rather well modelled and applied to the case of nuclear vessels (Benboudjema and Torrenti 2008) or buffers used for the storage of radioactive wastes (Craeye et al. 2009). Note that elevated temperatures at early age could also be the source of delayed ettringite formation. Cracking due to this reaction affects strongly the transfer properties of concrete (Al Shamaa et al. 2014).

After a long time, cracks are mainly due to drying shrinkage. An extensive literature is available on drying shrinkage and dimensional stability (de Sa et al. 2008; Bissonnette et al. 1999; Saito et al. 1991). As far as desiccation is involved, the permeability and the desorption isotherm and the couplings with cracking should be considered [see an example of model in (Torrenti and Benboudjema 2012)].

The cracking pattern and interconnectivity between micro and macrocracks have been reported to play an important role in the transport properties of the material (Lim et al. 2000). The development of microcracks facilitates the transport of aggressive salts and ions into the material, accelerating its deterioration. Most of the studies performed to evaluate the influence of cracking are based on the measurement of water and chloride permeability of pre-cracked specimens obtained from controlled compression or splitting tensile testing (Lim et al. 2000; Saito and Ishimori 1995; Djerbi et al. 2008; Wang et al. 1997; Ludirdja et al. 1989; Samaha and Hover 1992). A comprehensive review of the effect of mechanical stress on permeability of concrete is provided by Hoseini et al. (2009).

Recently, Rougelot et al. (2009) proposed a new method to investigate the effect of cracks on transport properties in cement pastes and mortars based on measurement of the extent of water vapour desorption and drying kinetics. Results demonstrated the influence of diffuse microcracking over macrocracks on the hydric patterns. Two methods of generating cracks in the material were used: (i) three points bending, leading to localised macrocracks and (ii) thermal shock, leading to more diffuse microcracking. Change in the compressive Young's modulus was used as an indicator of the damage induced by the thermal shock. Results showed that localised macrocracks did not influence the kinetics of water desorption while diffuse microcracking affected the transport properties of the material, most likely due to an increase in pore connectivity. The more connected porous network led to acceleration in the drying process and removal of water from ink-bottle pores. The Young's modulus after thermal shock was found to decrease by 15–20 % and was more significant for mortars than cement pastes. While these results clearly indicated the importance of considering the cracking pattern when dealing with durability issues, the rather severe treatment from thermal shock could have affected the paste mineralogy and differential shrinkage between the paste and the aggregate may have occurred.

Additionally, it should be mentioned that in the context of radioactive waste storage and disposal, concrete structures could be subjected to temperatures as high as 80 °C and thus to subsequent desiccation and cracking. The impact of temperature on the water movement and sorption properties of concrete has been poorly studied and results are scarce. Pihlajavaara (1976) had studied this effect and found a great impact of temperature and a reduction of the water content at saturation when temperature increases. Recent studies by Poyet (2009) and Poyet and Charles (2009) on desorption isotherms of CEM I HPC concrete performed at 30 and 80 °C confirmed these results. The isotherm shape was drastically modified and the water content at equilibrium was strongly reduced at 80 °C. The water content at saturation was significantly lower at 80 °C than at 30 °C. The description of the impact of temperature on concrete sorption properties was achieved using the Clausius-Clapeyron equation and the isosteric heat of sorption. Results obtained allowed for computation of concrete desorption isotherms at any other temperature with a good accuracy. Concrete durability assessment in such hydro-thermal environment should integrate this kind of approach.

Note that concrete carbonation should also influences the desorption isotherms (Auroy et al. 2012) and the transfer properties.

2.10 Understanding Durability and Performance in the Context of Specific Applications

Material performance specifications, formulations, and environmental stresses that challenge long-term durability are strongly a function of specific applications. Thus, there is need for general definition of performance envelopes (e.g., range of conditions and performance requirements) that are important for different applications. The resulting performance envelopes and key information gaps within each performance envelope provide an important framework for guiding future research. For example, reducing conditions and interfaces between cementitious wasteforms and concrete containment structures, are important considerations for grouted low activity waste disposal (Pabalan et al. 2012; Protiere et al. 2012), but not for many other structural applications. Also, specific waste types, such as low activity salt waste at the United States Savannah River Site and ion-exchange resins may impose unique material composition and performance challenges (Lafond et al. 2012; Neji et al. 2012; Pabalan et al. 2012; Protiere et al. 2012). Reactions and leaching with boric acid are of interest specifically for used fuel storage pools (Pabalan and Chiang 2012). Interactions of concrete at interfaces with bitumen, clays and salt formations, as well as bacterially mediated reactions, are important for some geologic repository designs (Fernández et al. 2009; Alquier et al. 2012; Bertron et al. 2012). However, understanding the impact of certain reactions and conditions on performance, such as carbonation (Atiş 2003; Khunthongkeaw and Tangtermsirikul 2005; Auroy et al. 2012; Brown et al. 2012a) and temperature effects (Kasami et al. 2012) have very broad applicability.

Additionally, keeping in mind operational aspects, it is essential to consider realistic degradation conditions by considering, for example, the water chemistry, the hydro-thermal environment (temperature/saturation), and the initial state of the material (e.g., carbonation, see recent work by Drouet et al. (2010) on the carbonation of blended cement-based materials).

2.11 Integration for Application to Engineering Systems

Integration of multiple phenomena and scaling to realistically-sized engineered systems under field conditions requires further research. For example, water flow by capillary and condensation processes and in response to temperature changes and thermal gradients is not often considered but was found to be central to understanding the performance of a field test case (Zuloaga et al. 2009). Drying shrinkage has been found to be dependent on the size of structural components (Benboudjema and Torrenti 2012). Denitrifying bacterial activity has also been observed at conditions anticipated for the cement matrix interface for wasteforms with high nitrate content, with important consequences on the understanding of radionuclide retention (Alquier et al. 2012). Probabilistic approach integrating the variability of key parameters controlling physic-chemical and mechanical properties is also a major issue for the long-term prediction of concretes structures stability and durability. These are just a few examples that point to the essential nature of large-scale field testing for validation of both conceptual and reactive transport models and the importance of long-term monitoring of test sites, with the intention of feedback to improve modelling and estimates of long-term performance.

2.12 RILEM TC-226-CNM Recommendations for Future R&D

Priorities for future R&D which should deserve further attention were highlighted during discussion panel sessions held at the end of the workshops:

- Multiple Processes and Properties have to be considered as a function of material composition
  - Processes: evolution of mineralogy and morphology, moisture transport, constituent transport, chemical reactions
  - Properties: thermal, hydraulic, mechanical, chemical properties
  - Constitutive relationships: link between processes and properties
- Need for guidance to establish consistent and complete data sets
- Need for better understanding of coupling (non-linear) of multiple processes and evolution of properties
- Need for better understanding of the interconnection pathways in cementitious materials with aggregates
- Need for better understanding of the effect of microstructure (including ITZ at the cement—aggregate interface) on effective diffusion and other transport properties

- Need to understand ageing and degradation phenomena for new materials (e.g., carbon fiber-cement paste interfaces in bulk material and exposure face, low pH cement)
- Need to be able to describe complex blended systems incorporating complex wasteforms
- Models need to take account of experimental data on real systems so that actual mineral phases and localised microstructure are not omitted
- It is essential that real systems (i.e. not idealised lab tests) need to be well characterised
- Extrapolating short-term data to the long-term is a major problem in a number of technical areas, including cement mineralogy and redox conditions
- The effect of microstructural changes on mechanical properties needs to be evaluated and the links between specific phase changes and mechanical changes needs to be demonstrated

References

Al Shamaa M., Lavaud, S., Divet, L., Nahas, G., Torrenti, J.M.: Coupling between mechanical and transfer properties and expansion due to DEF in a concrete of a nuclear power plant. Nucl. Eng. Des. **266**, 70–77 (2014)

Alquier, M., Kassim, C., et al.: Nitrate reducing bacterial activity in concrete cells of nuclear waste disposal. In: Proceedings of NUCPERF 2012 Workshop, Long-Term Performance of Cementitious Barriers and Reinforced Concrete in Nuclear Power Plant and Radioactive Waste Storage and Disposal Special issue of The European Physical Journal, vol. 56—2013 (2012)

Arnold, J., Meeussen, J.C.L., et al.: The impact of mineral deposition by carbonation and dissolution on tortuosity for leaching of constituents from a cementitious wasteform. In: L'Hostis, V., Gens, R., Gallé, C. (eds.) Proceedings of the RILEM Workshop on Long-Term Performance of Cementitious Barriers and Reinforced Concrete in Nuclear Power Plants and Waste Management, EFC Event 317, RILEM Proceedings PRO 64 (2009). ISBN: 978-2-31158-072-

Arnold, J., Kosson, D.S., et al.: Characterisation and modeling of major constituent equilibrium chemistry of a blended cement mortar. In: Proceedings of NUCPERF 2012 Workshop, Long-Term Performance of Cementitious Barriers and Reinforced Concrete in Nuclear Power Plant and Radioactive Waste Storage and Disposal Special issue of The European Physical Journal, vol. 56—2013 (2012)

Arnold, J., Kosson, D.S., et al.: Solution of the nonlinear Poisson-Boltzmann equation: application to ionic diffusion in cementitious materials. Cem. Concr. Res. **44**, 8–17 (2013)

Arns, C.H., Knackstedt, M.A., et al.: Virtual permeametry on microtomographic images. J. Petrol. Sci. Eng. **45**(1–2), 41–46 (2004)

Atiş, C.D.: Accelerated carbonation and testing of concrete made with fly ash. Constr. Build. Mater. **17**(3), 147–152 (2003)

Auroy, M., Poyet, S., et al.: Impact of carbonation on the durability of cementitious materials: water transport properties characterisation. In: Proceedings of NUCPERF 2012 Workshop, Long-Term Performance of Cementitious Barriers and Reinforced Concrete in Nuclear Power Plant and Radioactive Waste Storage and Disposal Special issue of The European Physical Journal, vol. 56—2013 (2012)

Banthia, N., Nandakumar, N.: Crack growth resistance of hybrid fiber reinforced cement composites. Cem. Concr. Compos. **25**(1), 3–9 (2003)

Banthia, N., Sheng, J.: Fracture toughness of micro-fiber reinforced cement composites. Cem. Concr. Compos. **18**(4), 251–269 (1996)

Benboudjema F., Torrenti J.M.: Early age behaviour of concrete nuclear containments. Nucl. Eng. Des. **238**(10), 2495–2506 (2008)

Benboudjema, F., Torrenti, J.-M.: Modelling desiccation shrinkage of large structures. In: Proceedings of NUCPERF 2012 Workshop, Long-Term Performance of Cementitious Barriers and Reinforced Concrete in Nuclear Power Plant and Radioactive Waste Storage and Disposal Special issue of The European Physical Journal, vol. 56—2013 (2012)

Bertron, A., Jacquemet, N., et al.: Physico-chemical interactions at the concrete-bitumen interface of nuclear waste repositories. In: Proceedings of NUCPERF 2012 Workshop, Long-Term Performance of Cementitious Barriers and Reinforced Concrete in Nuclear Power Plant and Radioactive Waste Storage and Disposal Special issue of The European Physical Journal, vol. 56—2013 (2012)

Bissonnette, B., Pierre, P., Pigeon, M.: Influence of key parameters on drying shrinkage of cementitious materials. Cem. Concr. Res. **29**(10), 1655–1662 (1999)

Briffaut, M., Benboudjema, F., Torrenti, J.M., Nahas, G.: A new analysis of the restrained ring shrinkage test. In: L'Hostis, V., Gens, R., Gallé, C. (eds.) In: Proceedings of the RILEM Workshop on Long-Term Performance of Cementitious Barriers and Reinforced Concrete in Nuclear Power Plants and Waste Management, EFC Event 317, RILEM Proceedings PRO 64 (2009). ISBN: 978-2-31158-072-

Brown, K. G., Arnold, J., et al.: Modeling Carbonation of High-Level Waste Tank Integrity and Closure. Proceedings of NUCPERF 2012 Workshop, Long-Term Performance of Cementitious Barriers and Reinforced Concrete in Nuclear Power Plant and Radioactive Waste Storage and Disposal Special issue of The European Physical Journal, vol. 56—2013 (2012a)

Brown, L., Sanchez, F., et al.: Performance of carbon nanofiber-cement composites subjected to accelerated decalcification. In: Proceedings of NUCPERF 2012 Workshop, Long-Term Performance of Cementitious Barriers and Reinforced Concrete in Nuclear Power Plant and Radioactive Waste Storage and Disposal Special issue of The European Physical Journal, vol. 56—2013 (2012b)

Castellote, M., Llorente, I., Andrade, C.: Influence of the composition of the binder and the carbonation on the zeta potential values of hardened cementitious materials. Cem. Concr. Res. **36**(10), 1915–1921 (2006)

Castellote, M., Andrade, C., Zuloaga, P., Ordoñez, M.: Chloride transport within the bulk of hardened cementitious materials. In: L'Hostis, V., Gens, R., Gallé, C. (eds.) Proceedings of the RILEM Workshop on Long-Term Performance of Cementitious Barriers and Reinforced Concrete in Nuclear Power Plants and Waste Management, EFC Event 317, RILEM Proceedings PRO 64. ISBN: 978-2-35158-072-1 (2009)

CBP: Review of Mechanistic Understanding and Modeling and Uncertainty Analysis Methods for Predicting Cementitious Barrier Performance (Contains 11 Chapters). CBP-TR-2009-002, Rev.0. Aiken, SC, Nashville, TN, Cementitious Barriers Partnership (2009)

Chatterji, S.: Transportation of ions through cement based materials. Part 2. Adaptation of the fundamental equations and relevant comments. Cem. Concr. Res. **24**(6), 1010–1014 (1994a)

Chatterji, S.: Transportation of ions through cement based materials. Part 3 experimental evidence for the basic equations and some important deductions. Cem. Concr. Res. **24**(7), 1229–1236 (1994b)

Chung, D.D.L.: Cement reinforced with short carbon fibers: a multifunctional material. Compos. B Eng. **31**(6–7), 511–526 (2000)

Chung, D.D.L.: Composites get smart. Mater. Today **5**(1), 30–35 (2002)

Colombet, P., Grimmer, A.-R., et al.: Nuclear Magnetic Resonance Spectroscopy of Cement-Based Materials. Springer, Berlin (1998)

Cong, X., Kirkpatrick, R.J.: 29Si MAS NMR study of the structure of calcium silicate hydrate. Adv. Cem. Based Mater. **3**(3–4), 144–156 (1996)

Craeye, B., De Schutter, G., Van Humbeeck, H., Wacquier, W., Van Cotthem, A.: Concrete buffers for the containment of high-level radioactive waste: casting conditions and early age behaviour, In: L'Hostis, V., Gens, R., Gallé, C. (eds.) Proceedings of the RILEM Workshop on Long-Term Performance of Cementitious Barriers and Reinforced Concrete in Nuclear Power Plants and Waste Management, EFC Event 317, RILEM Proceedings PRO 64. ISBN: 978-2-31158-072-

Dauzères, A., Le Bescop, P., et al.: Physico-chemical investigation of clayey/cement-based materials interaction in the context of geological waste disposal: Experimental approach and results. Cem. Concr. Res. **40**(8), 1327–1340 (2010a)

Dauzères, A., P. Le Bescop, et al.: Study of CEM I and low pH cement pastes leaching in multi-ionic underground water. In: 6th International Conference on Concrete under sever conditions - Environment and loading (CONSEC'10), 7–9 June 2010, Merida, Mexico (2010b)

de Larrard, T., Benboudjema, F., Colliat, J.B., Torrenti, J.M., Deleruyelle, F.: Correlations between measures and development of a predictive model for the long-time behaviour of concrete structures subjected to leaching. In: L'Hostis, V., Gens, R., Gallé, C. (eds.) Proceedings of the RILEM Workshop on Long-Term Performance of Cementitious Barriers and Reinforced Concrete in Nuclear Power Plants and Waste Management, EFC Event 317, RILEM Proceedings PRO 64. ISBN: 978-2-31158-072-

de Sa, C., Benboudjema, F., Thiery, M., Sicard, J.: Analysis of microcracking induced by differential drying shrinkage. Cem. Concr. Compos. **30**(10), 947–956 (2008)

Djerbi, A., Bonnet, S., Khelidj, A., Baroghel-bouny, V.: Influence of traversing crack on chloride diffusion into concrete. Cem. Concr. Res. **38**(6), 877–883 (2008)

Drouet, E., S. Poyet, et al.: Chemical changes and carbonation profiles of carbonated cement pastes at 80°C for different relative humidities. In: 6th International Conference on Concrete Under Sever Conditions—Environment and Loading (CONSEC'10), 7–9 June 2010, Merida, Mexico (2010)

Elakneswaran, Y., Nawa, T., et al.: An integrated model to predict the relations between electrical double layer properties and chloride ingress in cement-based materials. In: L'Hostis, V., Gens, R., Gallé, C. (eds.) Proceedings of the RILEM Workshop on Long-Term Performance of Cementitious Barriers and Reinforced Concrete in Nuclear Power Plants and Waste Management, EFC Event 317, RILEM Proceedings PRO 64 (2009). ISBN: 978-2-31158-072

Elakneswaran, Y., Iwasa, A., et al.: Ion-cement hydrate interactions govern multi-ionic transport model for cementitious materials. Cem. Concr. Res. **40**(12), 1756–1765 (2010)

Fernández, R., Cuevas, J., et al.: Modelling concrete interaction with a bentonite barrier. Eur. J. Mineral. **21**(1), 177–191 (2009)

Fu, X., Chung, D.D.L.: Contact electrical resistivity between cement and carbon fiber: its decrease with increasing bond strength and its increase during fiber pull-out. Cem. Concr. Res. **25**(7), 1391–1396 (1995)

Garboczi, E.J., Bentz, D.P., et al.: Digital images and computer modelling. In: Wong P.-Z. (eds.) Methods in the Physics of Porous Media, vol. 35, pp. 1–41. Academic Press, San Diego (1999)

Garcia Calvo, J. L., Hidalgo, A., et al.: Low pH concretes based on CAC for underground repositories of HLRW. Resistance respect to groundwater aggression. In: L'Hostis, V., Gens, R, Gallé, C. (eds.) Proceedings of the RILEM Workshop on Long-Term Performance of Cementitious Barriers and Reinforced Concrete in Nuclear Power Plants and Waste Management, EFC Event 317, RILEM Proceedings PRO 64 (2009). ISBN: 978-2-31158-072-

Garrabrants, A.C., Kosson, D.S., et al.: Background Information for the Leaching Environmental Assessment Framework (LEAF) Test Methods. EPA-600/R-10/170. U.S. Environmental Protection Agency, Office of Research and Development, Washington, DC (2010)

Garrabrants, A.C., Kosson, D.S., et al.: Interlaboratory Validation of the Leaching Environmental Assessment Framework (LEAF) Method 1314 and Method 1315. EPA/600/R-12/624. U.S. Environmental Protection Agency, Office of Research and Development, Washington, DC (2012a)

Garrabrants, A.C., Kosson, D.S., et al. Interlaboratory Validation of the Leaching Environmental Assessment (LEAF) Method 1313 and Method 1316. EPA-600/R-12/623. U.S. Environmental Protection Agency, Office of Research and Development, Washington, DC (2012b)

Han, S.H., Park, W.S.: Evaluation of carbonation-induced corrosion initiation for coastal concrete structure. In: L'Hostis, V., Gens, R., Gallé, C. (eds.) Proceedings of the RILEM Workshop on Long-Term Performance of Cementitious Barriers and Reinforced Concrete in Nuclear Power Plants and Waste Management, EFC Event 317, RILEM Proceedings PRO 64, ISBN: 978-2-35158-072-1 (2009)

Hidalgo López, A., García Calvo, J.L., et al.: Microstructural evolution of calcium aluminate cements hydration with silica fume and fly ash additions by scanning electron microscopy, and mid and near-infrared spectroscopy. J. Am. Ceram. Soc. **91**(4), 1258–1265 (2008)

Hoseini, M., Bindiganavile, V., Banthia, N.: The effect of mechanical stress on permeability of concrete: a review. Cem. Concr. Compos. **31**(4), 213–220 (2009)

Kasami, H., Tamura, M., et al.: Effects of cement type on the long-term performance of concrete exposed to sustained elevated temperatures. In: Proceedings of NUCPERF 2012 Workshop, Long-Term Performance of Cementitious Barriers and Reinforced Concrete in Nuclear Power Plant and Radioactive Waste Storage and Disposal Special issue of The European Physical Journal, vol. 56—2013 (2012)

Katz, A.: Effect of fiber modulus of elasticity on the long-term properties of micro-fiber reinforced cementitious composites. Cem. Concr. Compos. **18**(6), 389–399 (1996)

Katz, A., Bentur, A.: Mechanical properties and pore structure of carbon fiber reinforced cementitious composites. Cem. Concr. Res. **24**(2), 214–220 (1994)

Kawashima, S., Hou, P., et al.: Modification of cement-based materials with nanoparticles. Cem. Concr. Compos. **36**, 8–15 (2013)

Khunthongkeaw, J., Tangtermsirikul, S.: Model for simulating carbonation of fly ash concrete. J. Mater. Civ. Eng. **17**(5), 570–578 (2005)

Kinomura, K., Baroghel-Bouny, V.: Pore structure and transport properties of high-volume fly ash materials used for radioactive waste disposal facilities. In: L'Hostis, V., Gens, R., Gallé, C. (eds.) Proceedings of the RILEM Workshop on Long-Term Performance of Cementitious Barriers and Reinforced Concrete in Nuclear Power Plants and Waste Management, EFC Event 317, RILEM Proceedings PRO 64, ISBN: 978-2-35158-072-1 (2009)

Kosson, D.S., van der Sloot, H.A., et al.: An integrated framework for evaluating leaching in waste management and utilisation of secondary materials. Environ. Eng. Sci. **19**(3), 159–204 (2002)

Lafond, E., Cau Dit Coumes, C., et al.: Why should Portland cement (OPC) be blended with blast furnace slag (BFS) for cationic exchange resins encapsulation? In: Proceedings of NUCPERF 2012 Workshop, Long-Term Performance of Cementitious Barriers and Reinforced Concrete in Nuclear Power Plant and Radioactive Waste Storage and Disposal Special issue of The European Physical Journal, vol. 56—2013 (2012)

Lange, D.A., Ouyang, C., et al.: Behaviour of cement based matrices reinforced by randomly dispersed microfibers. Adv. Cem. Based Mater. **3**(1), 20–30 (1996)

Larbi, B., Dridi, W. et al.: Link between microstructure and tritiated water diffusivity in mortars. In: Proceedings of NUCPERF 2012 Workshop, Long-Term Performance of Cementitious Barriers and Reinforced Concrete in Nuclear Power Plant and Radioactive Waste Storage and Disposal Special issue of The European Physical Journal, vol. 56—2013 (2012)

Li, V.C.: On engineered cementitious composites (ECC)—a review of the material and its applications. J. Adv. Concr. Technol. **1**(3), 215–230 (2003)

Lim, C.C., Gowripalan, N., Sirivivatnanon, V.: Microcracking and chloride permeability of concrete under uniaxial compression. Cem. Concr. Compos. **22**(5), 353–360 (2000)

Lothenbach, B., Le Saout, G., et al.: Influence of limestone on the hydration of Portland cements. Cem. Concr. Res. **38**(6), 848–860 (2008)

Ludirdja, D., Berger, R.L., Young, J.F.: Simple method for measuring water permeability of concrete. ACI Mater. J. **86**(5), 433–439 (1989)

Martys, N.S., Hagedorn, J.G.: Multiscale modeling of fluid transport in heterogeneous materials using discrete Boltzmann methods. Mater. Struct. **35**(10), 650–658 (2002)

Mattigod, S.V., Lindberg, M.J., et al.: Waste Acceptance Testing of Secondary Wasteforms: Cast Stone, Ceramicrete and Duralith. PNNL-20632. Pacific Northwest National Laboratory, Richland, WA (2011)

Neji, M., Bary, B., et al.: Modelling of the interaction between chemical and mechanical behaviour of ion exchange resins incorporated into a cement-based matrix. In: Proceedings of NUCPERF 2012 Workshop, Long-Term Performance of Cementitious Barriers and Reinforced Concrete in Nuclear Power Plant and Radioactive Waste Storage and Disposal Special issue of The European Physical Journal, vol. 56—2013 (2012)

Nelson, P.K., Li, V.C., et al.: Fracture toughness of microfiber reinforced cement composites. J. Mater. Civ. Eng. **14**(5), 384–391 (2002)

Pabalan, R.T., Glasser, F.P., et al.: Review of Literature and Assessment of Factors Relevant to Performance of Grouted Systems for Radioactive Waste Disposal. CNWRA 2009-001. Center for Nuclear Waste Regulatory Analyses, San Antonio, TX (Prepared for the U. S. Nuclear Regulatory Commission Contract NRC NRC-02-07-006) (2009)

Pabalan, R.T., Alexander, G.W., et al.: Experimental study of contaminant release from reducing grout. In: Proceedings of NUCPERF 2012 Workshop, Long-Term Performance of Cementitious Barriers and Reinforced Concrete in Nuclear Power Plant and Radioactive Waste Storage and Disposal Special issue of The European Physical Journal, vol. 56—2013 (2012a)

Pabalan, R.T., Chiang, K.-T. K.: Experimental study and reactive transport modeling of boric acid leaching of concrete. Proceedings of NUCPERF 2012 Workshop, Long-Term Performance of Cementitious Barriers and Reinforced Concrete in Nuclear Power Plant and Radioactive Waste Storage and Disposal Special issue of The European Physical Journal, vol. 56—2013 (2012b)

Pacheco-Torgal, F., Miraldo, S., et al.: Targeting HPC with the help of nanoparticles: an overview. Constr. Build. Mater. **38**, 365–370 (2013)

Pihlajavaara, 1976, On practical estimation of moisture content of drying concrete; Il Cemento, n° 3, 1976

Poyet, S.: Experimental investigation of the effect of temperature on the first desorption isotherm of concrete. Cem. Concr. Res. **39**(11), 1052–1059 (2009)

Poyet, S., Charles, S.: Temperature dependence of the sorption isotherms of cement-based materials: heat of sorption and Clausius-Clapeyron formula. Cem. Concr. Res. **39**(11), 1060–1067 (2009)

Protiere, Y., Samson, E., et al.: Modeling the long-term durability of concrete barriers in the context of low-activity waste storage. In: Proceedings of NUCPERF 2012 Workshop, Long-Term Performance of Cementitious Barriers and Reinforced Concrete in Nuclear Power Plant and Radioactive Waste Storage and Disposal Special issue of The European Physical Journal, vol. 56—2013 (2012)

Raki, L., Beaudoin, J.J., et al.: Cement and concrete nanoscience and nanotechnology. Materials **3**, 918–942 (2010)

Ramachandran, V.S., Beaudoin, J.J.: Handbook of Analytical Techniques in Concrete Science and Technology, William Andrew Publishing/Noyes. Available at: http://www.knovel.com/web/portal/browse/display?_EXT_KNOVEL_DISPLAY_bookid=265&VerticalID=0 (2001)

Revil, A.: Ionic diffusivity, electrical conductivity, membrane and thermoelectric potentials in colloids and granular porous media: a unified model. J. Colloid Interface Sci. **212**(2), 503–522 (1999)

Reza, F., Batson, G.B., et al.: Resistance changes during compression of carbon fiber cement composites. J. Mater. Civil Eng. ASCE **15**, 476–483 (2003)

Rougelot, T., Burlion, N., Skoczylas, F.: Influence of pre-cracking over water desorption in cement pastes and mortars. In: RILEM PRO 64 Proceedings of the NUCPERF 2009 Workshop (Long-Term Performance of Cementitious Barriers and Reinforced Concrete in Nuclear Power Plants and Waste Management, pp. 125–133 (EFC Event n°317). Cadarache, France (2009)

Saito, M., Ishimori, H.: Chloride permeability of concrete under static and repeated compressive loading. Cem. Concr. Res. **25**(4), 803–808 (1995)

Saito, M., Kawamura, M., Arakawa, S.: Role of aggregate in the shrinkage of ordinary portland and expansive cement concrete. Cem. Concr. Compos. **13**(2), 115–121 (1991)

Samaha, H.R., Hover, K.C.: Influence of microcracking on the mass transport properties of concrete. ACI Mater. J. **89**(4), 416–424 (1992)

Sanchez, F., Borwankar, A.: Multi-scale performance of carbon microfiber reinforced cement-based composites exposed to a decalcifying environment. Mater. Sci. Eng. A **527** (13–14), 3151–3158 (2010)

Sanchez, F., Borwankar, A., et al.: Effect of decalcification on the performance of carbon microfiber reinforced cement-based materials. In: L'Hostis, V., Gens, R., Gallé, C. (eds.) Proceedings of the RILEM Workshop on Long-Term Performance of Cementitious Barriers and Reinforced Concrete in Nuclear Power Plants and Waste Management, EFC Event 317, RILEM Proceedings PRO 64 (2009). ISBN: 978-2-31158-072-

Sanchez, F., Sobolev, K.: Nanotechnology in concrete: a review. Constr. Build. Mater. **24**(11), 2060–2071 (2010)

Sarkar, S., Mahadevan, S., et al.: Numerical simulation of cementitious materials degradation under external sulfate attack. Cem. Concr. Compos. **32**(3), 241–252 (2010)

Sarkar, S., Mahadevan, S., et al.: Sensitivity analysis of damage in cement materials under sulfate attack and calcium leaching. J. Mater. Civ. Eng. **24**(4), 430–440 (2012)

Shah, S.P., Kuder, K.G., et al.: Fiber-reinforced cement based composites: a forty year odyssey. In: Prisco, M.D., Felicetti, R., Plizzari, G.A. (eds.) 6th RILEM Symposium on Fibre-Reinforced Concretes (FRC), p. 28, RILEM Publications SARL, Varenna, Italy (2004)

Snyder, K., Stutzman, P.E., et al.: Hydrated phases in blended cementitious systems for nuclear infrastructure. In: L'Hostis, V., Gens, R., Gallé, C. (eds.) Proceedings of the RILEM Workshop on Long-Term Performance of Cementitious Barriers and Reinforced Concrete in Nuclear Power Plants and Waste Management, EFC Event 317, RILEM Proceedings PRO 64 (2009). ISBN: 978-2-31158-072-

Stora, E., Conrardy, C., Barbarulo, R., Chen, J., Thoué, F.: Investigation of the atmospheric carbonation mechanisms of cementitious materials through numerical modeling. In: L'Hostis, V., Gens, R., Gallé, C. (eds.) Proceedings of the RILEM Workshop on Long-Term Performance of Cementitious Barriers and Reinforced Concrete in Nuclear Power Plants and Waste Management, EFC Event 317, RILEM Proceedings PRO 64 (2009). ISBN: 978-2-31158-072-

Thouvenot, P., Bildstein, O., et al.: Numerical simulations of atmospneric carbonation in concrete components of a deep geological intermediate-low level waste disposal. In: Proceedings of NUCPERF 2012 Workshop, Long-Term Performance of Cementitious Barriers and Reinforced Concrete in Nuclear Power Plant and Radioactive Waste Storage and Disposal Special issue of The European Physical Journal, vol. 56—2013 (2012)

Torrenti, J.M., Benboudjema, F.: Modelling desiccation shrinkage of large structures. In: Proceedings of NUCPERF 2012 Workshop, Long-Term Performance of Cementitious Barriers and Reinforced Concrete in Nuclear Power Plant and Radioactive Waste Storage and Disposal Special issue of The European Physical Journal, vol. 56—2013 (2012)

Torrents, J.M., Mason, T.O., et al.: Impedance spectra of fiber-reinforced cement-based composites: a modeling approach. Cem. Concr. Res. **30**(4), 585–592 (2000)

Toutanji, H.A., El-Korchi, T., et al.: Strength and reliability of carbon-fiber-reinforced cement composites. Cem. Concr. Compos. **16**(1), 15–21 (1994)

Ukrainczyk, N., Koenders, E.A.B., et al.: Numerical estimation of transport properties of cementitious materials using 3D digital images. In: Proceedings of NUCPERF 2012 Workshop, Long-Term Performance of Cementitious Barriers and Reinforced Concrete in Nuclear Power Plant and Radioactive Waste Storage and Disposal Special issue of The European Physical Journal, vol. 56—2013 (2012)

Um, W., Serne, R.J., et al.: Radionuclide Retention Mechanisms in Secondary Waste-Form Testing, Phase II. PNNL-20753. Pacific Northwest National Laboratory, Richland, WA (2011)

USEPA: Test Methods for Evaluating Solid Waste, Physical/Chemical Methods (SW-846) New Methods Page. See http://www.epa.gov/waste/hazard/testmethods/sw846/ (2013)

Wang, Y., Backer, S., et al.: An experimental study of synthetic fibre reinforced cementitious composites. J. Mater. Sci. **22**, 4281–4291 (1987)
Wang, K., Jansen, D.C., Shah, S.P., Karr, A.F.: Permeability study of cracked concrete. Cem. Concr. Res. **27**(3), 381–393 (1997)
Weiss, J. (2012). Transport Through Cracked Porous Media with Emphasis on Concrete. Attachment 1 to Transport Through Cracked Concrete: Literature Review. SRNL-STI-2012-00267, Rev. 0. Savannah River National Laboratory, Aiken, SC
Yu, S.W., Page, C.L.: Diffusion in cementitious materials: 1. Comparative study of chloride and oxygen diffusion in hydrated cement pastes. Cem. Concr. Res. **21**(4), 581–588 (1991)
Zuloaga, P., Andrade, C., et al.: Leaching of reinforced concrete vaults subjected to variable water content due to capillary suction created by seasonal temperature changes. In: L'Hostis, V., Gens, R., Gallé, C. (eds.) Proceedings of the RILEM Workshop on Long-Term Performance of Cementitious Barriers and Reinforced Concrete in Nuclear Power Plants and Waste Management, EFC Event 317 (2009). RILEM Proceedings PRO 64, ISBN: 978-2-31158-072-

Chapter 3
Corrosion

Bruno Kursten and Nick Smart

Abstract The corrosion of steel in cementitious environments in nuclear applications has been studied extensively because (i) steel is widely used as reinforcement in the concrete used in nuclear power plants (NPPs) and (ii) cementitious materials are used as a waste encapsulant or backfill in some disposal concepts for low level and high level radioactive waste, where steel waste containers are used and steel wastes are processed. This section summarises the papers on corrosion in concrete that were given at the NUCPERF series of conferences. It first presents an overview of potential corrosion issues in NPPs and waste facilities and then describes the main corrosion mechanisms for steel in concrete (including passive corrosion, chloride–induced corrosion and carbonation). This is followed by an overview of the various aspects of modelling of corrosion of steel in concrete that were discussed during the NUCPERF meetings. The features of corrosion in concrete that need to be taken into consideration when developing new test programmes are summarised and the possible effects of concrete cracking on corrosion behaviour are discussed. Finally, some recommendations for future R&D on the subject of corrosion in concrete in nuclear applications are made.

Keywords Metallic radioactive wastes · Steel reinforcements · Chloride-induced-corrosion · Carbonation-induced-corrosion · Modeling

3.1 Introduction

Reinforced concrete is a structural material that is widely used in nuclear power plant and therefore the corrosion behaviour of steel reinforcement in concrete is of major importance to the safe operation of nuclear power plant (NPP). In addition,

B. Kursten
SCK-CEN, Mol, Belgium

N. Smart (✉)
AMEC, Didcot, UK
e-mail: nick.smart@amec.com

V. L'Hostis and R. Gens (eds.), *Performance Assessment of Concrete Structures and Engineered Barriers for Nuclear Applications*,
RILEM State-of-the-Art Reports 21, DOI 10.1007/978-94-024-0904-8_3

cement-based materials are widely used in the packaging and disposal of radioactive waste, from low-level waste (LLW) to high level waste (HLW), and the corrosion of carbon steel, stainless steel and other waste metals in cementitious environments is therefore an area of significant interest to the nuclear industry. NPP operations and radioactive waste management are the two main areas relating to corrosion issues that were covered during the workshops organised by the RILEM TC 226 technical committee over the period 2009–2012. The purpose of this chapter is to summarise the key points and themes regarding corrosion that arose during the workshops. The main aspects of corrosion that were covered by papers presented at the workshops can be broadly broken down into the following:

- Corrosion of reinforcement in nuclear power plant
- Corrosion during radioactive waste management and disposal
- Passive corrosion
- Chloride-induced corrosion in concrete
- Corrosion in carbonated concrete
- Corrosion modelling
- In situ experiments and natural analogues
- Corrosion test programmes
- Prevention of corrosion in concrete
- Mechanical properties of corrosion products
- Corrosion monitoring

An overview of the themes, key points and conclusions relating to corrosion that arose from the papers presented in these sessions, and which are published in the conference proceedings, is given in the following sections. The papers presented are grouped together for the various themes that are discussed, rather than separated into discussion about the individual workshops, but the reference list shows the workshop at which the papers were presented. The final section contains a list of recommendations for further research in the field of corrosion in concrete structures in nuclear applications, based on the discussions at the workshops.

Cementitious materials provide protection to steels because of the highly alkaline nature of the water in the pores of the cement, which passivates the surface of steels and hence controls the corrosion rate at very low values. However, the protection offered by the passive film can be disrupted by the presence of aggressive species, particularly chloride, and degradation of the concrete, for example by carbonation, which reduces the pH of the porewater in the vicinity of the metal surface. The performance and ageing behaviour of the passive film formed on the steel surface in alkaline media over the extended timescales required in the nuclear industry, which span a few decades in the case of nuclear power plant applications, to thousands of years in the case of radioactive waste disposal, is of great interest. Understanding the behaviour of the passive films on metal surfaces in alkaline conditions and concrete is fundamental to the safe management of nuclear facilities that employ cementitious materials in conjunction with metal structures. Such an understanding is important for a number of important reasons, including underpinning safety analysis, operational and maintenance planning, lifetime assessment, economic forecasting, etc. Extensive literature exists on the corrosion of steel in concrete [for example in references (Bertolini et al. 2004; Bentur 1997; Broomfield 2007; Ahmad 2006; Mindess et al. 2003)] and the reader is directed towards these texts for a discussion of the principles of corrosion of metals, particularly steel in concrete. The following sections summarise the main topics that were discussed at the various workshops organised by TC-226.

3.2 Corrosion of Metals in Concrete for Nuclear Applications

The main applications for metals in concrete in nuclear applications are for the reinforcement in nuclear power plant (NPP) and as a waste encapsulant for radioactive waste disposal. It is also a potential backfill in some geological disposal facility concepts. A schematic summary of the potential corrosion issues for these two main applications is shown in Fig. 3.1.

The study of corrosion of steel reinforced structures in NPPs differs from that in interim storage and disposal facilities in the sense that:

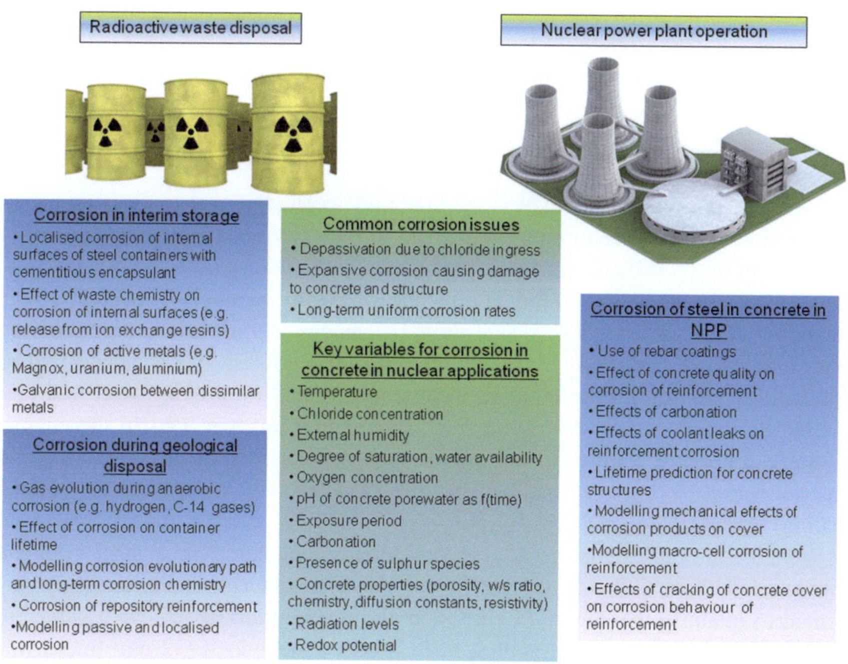

Fig. 3.1 Schematic summary of corrosion issues affecting metals in cement-containing nuclear applications

- Different timeframes apply (<100 years for NPP; >1000 years for waste facilities)
- Different environmental conditions apply (in NPP: atmospheric corrosion due to carbonation; in waste disposal facilities: conditions evolve from oxic to anoxic)
- Different maintenance possibilities apply (in NPP: little effort on studying the corrosion mechanisms, but more emphasis is put on repair techniques; in waste disposal facilities: more emphasis is put on long-term integrity, because repair is rather difficult)

3.2.1 *Steel Reinforced Structures in NPPs*

Corrosion problems in NPPs can be located in:

- the containment building of the reactor
- cooling towers
- cooling pipes made of concrete with a steel liner, close to seawater (corrosion due to ingress of chlorides)

Wegemar (2006) pointed out the importance of adopting detailed work instructions and adequate quality control throughout the construction works in order to ensure long-term performance for facilities with embedded steel liners or reinforcements. Penetrations of a BWR embedded reactor containment liner (carbon steel, 7 mm thick) were observed near a feedthrough for electrical conductors. Porous concrete was used and cavities filled with air had formed during construction. It was concluded that anodic dissolution of the steel liner has occurred as a result of an insufficient passivation of the surface of the carbon steel liner. In addition, the corrosion process was initiated and accelerated due to the presence of increased chloride concentrations inside the formed cavities (depassivation).

Weiss et al. (2009) studied the use of coatings to address both the bonding and the corrosion problem in reinforced concrete structural members of nuclear power plants. An innovative alkali-resistant vitreous enamel coating (based on zirconium oxide) was developed. They concluded that the application of this coating could be beneficial in improving the bond between concrete and reinforcing steel and preventing corrosion of the underlying steel. Even small imperfections in the enamel coating did not pose a threat for the corrosion performance of the system (corrosion was not found to propagate neither into the steel bar, nor underneath the enamel coating).

Brem et al. (2009) discussed the problems encountered in a steel concrete reinforced wall of a Swiss nuclear power plant, together with the analysis performed to quantify the degree of corrosion and the solution to achieve a sufficient service life of the concrete structure. In this case, a steel shell (32 mm thick) was attached to a thick reinforced concrete wall, which were separated by a 20 mm joint. The joint between the concrete and the steel structure was filled with rock wool. Inspection of the steel structure revealed atmospheric corrosion in the upper part and localised corrosion in the lower part of the joint. Regular flooding of the joint with water containing boric acid decreased the alkalinity of the concrete at the interface of the concrete and the steel structure leading to acid attack of the steel. A cathodic protection system was installed with the aim to reduce the corrosion rate. After reparation of the joint sealing, an on-line corrosion monitoring system was incorporated that enabled to measure a significant reduction of the corrosion rate.

Pabalan et al. considered the effect of boric acid on the corrosion of steel in concrete structures, related to spent fuel pools containing borated water. He used (i) modelling of the interactions between boric acid and concrete (Pabalan and Chiang 2013) and (ii) measuring the corrosion behaviour of carbon steel reinforcement in test environments that simulated the concrete environment that had reacted with boric acid and concrete (Pabalan et al. 2013). pH profiles as a function of time were calculated for a range of parameters and it was found that the corrosion rate of rebars increased abruptly when the pH in the vicinity of the steel was lower than 6.8–7.3. The assessment of corrosion of rebar in ageing management programmes is discussed in Chap. 4.

3.2.2 *Steel Reinforced Structures in Waste Facilities (Interim Storage, Underground Repository)*

Discussion of the corrosion issues for steel in cementitious environments during waste management can be divided into the following main phases, for all radioactivity levels for the waste (i.e. LLW to HLW):

- Interim surface storage after packaging the waste
- After placement in an underground repository
- After closure of the repository

The detailed arrangements will depend on the activity level and the national strategy for radioactive waste management. Metal corrosion affects the management of radioactive waste in a number of key areas, including

- Corrosion controls the lifetime of the metallic waste containment
- Corrosion controls the generation of gases from packaged waste
- Corrosion can generate expansive forces, which could disrupt the waste packages and lead to cracking of the concrete matrix

The papers given at the workshops on the subject of corrosion during radioactive waste disposal are summarised below.

Metallic radioactive waste containers during interim surface storage

The internal chemistry of wastes encapsulated in cement-based grouts can affect the corrosion of both carbon steel and stainless steel waste containers. For example, aggressive species, such as chlorides or sulphate, released from ion exchange resins or other polymers in a radiation field may be corrosive (Farina et al. 2009; Farina 2013; Fennell 2009).

Metallic radioactive waste containers in underground repositories

There was a general consensus at the workshops that the pH will remain very high for a long period in OPC-based environments (pH $\geq$ 12.5). The alkaline environment will maintain steel in a passive state (i.e. at a very low corrosion rate). When attempting to predict the long-term corrosion damage in nuclear waste systems it is necessary to define the corrosion evolutionary path (CEP) over hundreds of thousands of years, that is how the environment in which corrosion is occurring changes with time, but it is very difficult to achieve this. The CEP can be modelled, but it is difficult to validate the results. Modelling the CEP is complicated by the coupling of processes in the long-term. There is a need to couple corrosion, radiation and geochemical models.

In the long-term, the environment in an underground repository will become anoxic due to consumption of oxygen by reaction with minerals and due to microbial activity and so the corrosion of steel will be subject to anaerobic corrosion processes. In relation to the Belgium Supercontainer Concept, a series of papers (Smart et al. 2009, 2011, 2013; Winsley et al. 2011) presented the results of measurements of the rate of anaerobic corrosion of carbon steel in alkaline media. The work studied the effects of temperature (25 and 80 °C) and radiation (25 Gy hr^{-1}) on the anaerobic corrosion behaviour, in the presence of chloride. The experiments used hydrogen evolution and electrochemical methods to measure the corrosion rate. Under anoxic conditions, the corrosion rate of carbon steel was found to be not highly dependent on the concentration of chloride and radiation dose (up to 25 Gy hr^{-1}). After several years' exposure the long-term corrosion rate was found to be <0.1 μm yr^{-1} (and still decreasing).

Jung et al. (2011) conducted a series of electrochemical experiments to investigate the effect of dissolved oxygen, pH, and chloride concentration on the corrosion rate of reinforcing steel in a geological disposal facility saturated with groundwater. It was found that the corrosion rate was proportional to the concentration of chloride and dissolved oxygen. The pH also strongly influenced the corrosion rate of the reinforcing steel. At the pH expected in the pore water of a concrete structure of a disposal facility (i.e. pH 10–12) and dissolved oxygen concentration of 1 mg L^{-1}, the corrosion rate of reinforcing steel was determined to be in the range of 0.01–0.001 μm yr^{-1}. The corrosion rates were higher than those estimated from an empirical model based on the diffusion of dissolved oxygen.

The combined effect of chloride and reduced oxygen content on the corrosion behaviour of steel in concrete was considered by Andrade et al. (2013). It was demonstrated that corrosion can occur even with a reduced oxygen content provided that high enough chloride concentration is present.

King and Padovani published a review of candidate materials for use as containers for HLW and/or spent fuel in the UK (King and Padovani 2011), including the option of placing the containers in cementitious backfill environments. The high-level review included consideration of carbon and stainless steels, nickel alloys and titanium as candidate canister materials.

Steel reinforcements in other underground facilities

Monitoring the corrosion processes has gained in importance in long-term corrosion predictions. A number of authors reported the results of investigations of the corrosion of reinforcement bar in underground waste management facilities:

- Duffó et al. (2006, 2013), Schulz et al. (2009) reported on the corrosion behaviour of steels involved in Intermediate Level Radioactive Waste Disposal (ILRWD) in Argentina. The corrosion rate decreased in time from 30 to 2 $\mu m\ yr^{-1}$ after 700 days. Correlation of data obtained in the laboratory with those obtained from field measurements was not always straightforward: the field measurements showed lower corrosion potentials and higher corrosion rates than those measured under laboratory conditions. The explanation for this discrepancy was that the laboratory specimens were free of rust, while the rebars used in the field measurements were covered with an air-formed corrosion layer
- Andrade et al. (2011) reported on a long-term monitoring programme of the El Cabril LLW/ILW disposal facility in Spain. The corrosion potential and corrosion rate of a dummy radioactive waste container embedded in the facility was monitored from 1994 onwards, together with various environmental parameters including temperature, concrete resistivity, strain and oxygen availability. Temperature was found to be the most significant factor determining the corrosion behaviour of the reinforcement in the dummy container

3.3 Corrosion Mechanisms

During the various workshops a number of different aspects of the corrosion mechanisms affecting metals in cement-based nuclear applications were discussed. These are summarised below.

3.3.1 Passive Corrosion

A consensus exists that in OPC-based environments, the pH will remain very high for a long period of time thereby maintaining the reinforcement steel in a passive state (i.e. at a very low corrosion rate) (Gens et al. 2006; Pourbaix and L'Hostis 2006; Kursten et al. 2011; L'Hostis et al. 2011).

A detailed analysis of corrosion product films and their electronic properties is required so that their behaviour over long time periods can be modelled, for example using the Point Defect Model or the high-field film growth model.

3.3.2 *Chloride-Induced Corrosion in Concrete*

Chloride ions can cause a local breakdown of the protective oxide film formed on the steel reinforcement in concrete, which results in a localised corrosion attack. The mechanism by which chlorides are able to destroy the protective film is yet not fully understood (ACI 2001; Mendoza 2003):

- One hypothesis is that the chloride ions become incorporated into the passive film and reduce its resistance
- A second hypothesis states that chloride ions are adsorbed on the metal surface in competition with dissolved oxygen or hydroxyl ions
- A third hypothesis postulates that the chloride ions compete with the hydroxyl ions for combining with Fe^{2+} cations (produced by corrosion) to form soluble complexes of iron chloride which can diffuse away from the anode destroying the protective oxide film and stimulating further dissolution of the metal

Irrespective of the mechanism, the net result is that active corrosion occurs at regions of local breakdown of the protective film (i.e. pits) and, once started, it proceeds autocatalytically, that is, in a self-generating manner. The chloride and ferrous ions react to form soluble complexes of iron chloride ($(FeCl_n)^{(2-n)}$, n = 1–4) (Pabalan et al. 2009), which can diffuse away from the anodic sites. When the complexes reach a region of high pH (i.e. the concrete) it breaks down, precipitating an insoluble iron hydroxide and liberating the Cl^- ions, which are then able to migrate back to the anodic sites (i.e. inside the pits) because chlorides, being negatively charged, are attracted to the anodic regions. The chloride ions stimulate further dissolution of the metal. In this overall process, hydroxyl ions are continuously consumed, locally decreasing the pH (i.e. making the solution inside the pits more and more acidic) and, thereby, also enhancing further metal dissolution (ACI 2001; Mendoza 2003; Hansson et al. 2007).

It is commonly accepted (Angst et al. 2009) that corrosion in non-carbonated, alkaline concrete can only take place when the chloride content at the surface of the reinforcement reaches a 'threshold value', which is often referred to in the literature as 'the critical chloride content'. The critical chloride content can be expressed as (i) the total chloride content relative to the weight of the cement, (ii) the free chloride content, either related to the weight of cement or concrete or as a concentration (mol L^{-1}) in the pore solution, or (iii) the ratio of the chloride ion activity to the pH of the pore solution, i.e. the $(Cl^-)/(OH^-)$ ratio (Angst et al. 2009).

There are analogies between atmospheric corrosion and corrosion in concrete with a limited water content (Charles et al. 2009). These analogies are probably more important in assessing the lifetime of concrete structures such as buildings rather than waste disposal facilities.

A combination of stainless steel or Alloy 22 and cementitious environments is good for resisting chloride attack. Legat et al. (2009) examined the corrosion behaviour of a wide variety of steel reinforcement (black steel, ferritic stainless steels, austenitic stainless steels, and duplex stainless steels) exposed to synthetic ground water or embedded in different types of concrete that were representative for the future LILW repository environment in Slovenia. They reported a ranking of the corrosion performance of the various types of steel in synthetic ground water: black steel < ferritic SS < austenitic SS < duplex SS. Tests performed in concrete cracked specimens revealed that corrosion activity was only measured on rebars made from black steel and ferritic stainless steel, while the austenitic and duplex stainless steel types remained in the passive state (Charles et al. 2009).

3.3.3 *Carbonation*

Carbonation does not cause any damage to the concrete itself. However, carbonation can lead to a lowering of the pH of the pore solution caused by a drastic reduction in the concentration of hydroxyl ions due to the following reaction (Bertolini et al. 2004)

$$CO_2 + Ca(OH)_2 \rightarrow CaCO_3 + H_2O. \tag{3.1}$$

The fall in pH of the concrete pore solution to below a certain threshold level may cause loss of passivity of the steel surface and initiate corrosion of the reinforcement embedded in concrete. Different values, ranging from 8.5 to 11.5, have been reported in the literature to define this pH threshold level (Hurley and Scully 2002).

Various aspects of corrosion of metals in concrete are important in nuclear applications, including radioactive waste container lifetime predictions (where the requirements are different for different disposal concepts), such as expansion due to corrosion processes (mechanical property data are needed for oxide films), and determination of hydrogen production rate due to anaerobic corrosion.

Dang et al. (2013) investigated the effect of cracks on the initiation and propagation of rebar corrosion induced by CO_2. Steel rebars were embedded in concrete that was mechanically cracked (crack widths ranged from 52 to ~400 μm). The cracked samples were kept in a CO_2 atmosphere at 23 °C and 65 % relative humidity for 23 weeks. They were then exposed to wetting-drying cycles (1 day under tap water—3 days in lab air). All interfaces were found to have been carbonated, even in case of the very fine crack widths. The maximum thickness of the rust layer was measured to be 86 μm.

3.4 Corrosion Modelling

During the workshops organised by the technical committee a number of papers dealt with modelling various aspects of the corrosion of metals in concrete, during their use in nuclear applications. Many aspects of corrosion have been modelled in the general literature but the current summary only deals with the papers presented at the workshops.

Modelling approaches have been used for representing (i) the mechanical effects of expansive corrosion products (but realistic mechanical property data are needed for the corrosion products) (François et al. 2006), (ii) the electrochemistry of film formation on metals (e.g. the Point Defect Model, PDM) (Macdonald et al. 2009, 2011a, b; Saleh 2011; Bataillon et al. 2010), (iii) water transport processes in concrete as a function of humidity or during wet-dry cycles and (iv) pitting corrosion.

The pitting factor is often used in assessing corrosion allowances for container materials. However, its use is based on experimental observation and it lacks a good mechanistic understanding.

Harnisch et al. (2009) used a combined experimental-modelling approach to study the macro-cell corrosion of rebar in concrete and in particular to predict the corrosion potential and corrosion current distribution around corroding reinforcement bar.

3.5 Corrosion Test Programmes

It is good practice to use a number of different independent corrosion rate measurement techniques (e.g. weight loss, electrochemical methods such as AC impedance and linear polarisation resistance, gas evolution methods) to increase confidence in the corrosion rate data. Also, analytical techniques can provide important supporting information about corrosion products (e.g. Raman spectroscopy). Finally, there is a need for a good statistical basis for key corrosion data used in repository performance assessments.

Important aspects to consider in designing/developing R&D programmes are related to:

- Environmental conditions
- Experimental parameters
- Test duration

Environmental conditions

When attempting to predict the long-term corrosion damage in nuclear waste systems, it is necessary to define the Corrosion Evolutionary Path (CEP, i.e. how the environment in which corrosion is occurring changes with time). A wide range of techniques exists to characterise the environment in which experiments may be conducted (e.g. temperature, relative humidity, resistivity, moisture availability in the concrete structure, possibility of differential aeration cells during the transient stage from oxic to anoxic). The expected CEP can be used to guide the definition of the experimental programme. The CEP can be modelled, but it is difficult to validate the results. Furthermore, modelling the CEP is complicated by the coupling of processes in the long-term.

A lot of experiments are carried out in solutions to represent cementitious pore water compositions, but there is also a need to carry out tests in solid matrices to allow for mass transport limitations. In addition, there may also be different behaviour at the interface between the metal and the surrounding solid matrix. The results from a range of experiments indicate that tests in solution give similar results to those in solid matrices.

There is a need to use a range of techniques to characterise the environment in which experiments will be conducted (e.g. T, RH, resistivity, moisture availability in the concrete structure, possibility of differential aeration cells during the transient stage from oxic to anoxic). These measurements can be used to define the CEP.

Experimental parameters

Important parameters to take into account include the initial state of the surface (e.g. pre-corroded, presence of 'as received' film), composition and microstructure of the metal and also de-bonding and the presence of voids and pores in the concrete.

Test duration

A general trend is observed of the corrosion rate decreasing with time. This is also observed in other systems (e.g. carbon steel exposed to a clay environment), including aerated and de-aerated systems. Test duration can play an important role in lifetime predictions of reinforced concrete structures; a proper estimation should be made of how long experiments should run in order to be able to extrapolate data to extended timescales.

In situ *experiments and analogues*

There is a need for in situ experiments and analogue investigations, such as archaeological artefacts, as well as laboratory experiments, to ensure that all possible failure mechanisms are taken into account (e.g. microbial effects). Moreover, these analogues can bring information regarding the evolution of the structure of corrosion product films and their transport properties (Chitty et al. 2006; Saheb et al. 2011).

3.6 Effects of Concrete Cracking on Corrosion Behaviour

At the NUCPERF 2012 workshop Professor Mark Alexander gave a keynote lecture (Alexander et al. 2012), (not published in the proceedings), which highlighted the important effect that cracking of the concrete cover has on the corrosion susceptibility, because cracks influence the rates of ingress of corrosion-inducing species, the nature of transport mechanisms and the corrosion kinetics. The other key factors controlling the corrosion rate are the resistivity of the concrete and cover depth. Alexander believes that it is not possible to define a threshold crack width for corrosion in concrete and that new corrosion rate prediction models should incorporate (directly or indirectly) the influence of cracking on corrosion rate, based on long-term data from in situ measurements, since there can be large discrepancies between laboratory and field measurements.

There is a need to characterise the mechanical properties of corrosion products, particularly when assessing the risk of expansive corrosion. These properties will evolve with time. The corrosion products may initially be in the form of gels, which develop into solid structures as the water content changes (e.g. due to water consumption by anaerobic corrosion) (Miserque et al. 2006; François et al. 2009; Caré et al. 2009; Neff et al. 2009).

3.7 RILEM TC-226-CNM Recommendations for Future R&D

Several knowledge gaps and needs for future research regarding corrosion in concrete for nuclear applications were identified during the operation of the TC 226 technical committee. These areas were generally highlighted during discussion panel sessions held at the end of the workshops and were compiled by the chairs of the corrosion sessions. The highlighted areas are summarised below.

1. Increased knowledge of many aspects related to the initiation and propagation of corrosion in these environments is needed. The depassivation due to a lack of oxygen, by pH lowering or by leaching still has to be better quantified.
2. It is known that diffusion of chlorides into concrete and carbonation lead to reinforcement corrosion, but studies of the evolution of the corrosion morphology and the corrosion rate, particularly in buried conditions, are scarce.
3. There is a need to perform experiments under realistic conditions, which accurately reflect the expected Corrosion Evolutionary Path (CEP), for both in situ and laboratory experiments. The CEP can be used as a guideline to define the experimental conditions.
4. There is a need for in situ experiments and natural analogue investigations, as well as laboratory experiments, to ensure that all possible failure mechanisms are taken into account (e.g. microbial effects). Such investigations can also

provide useful information regarding the transport properties of corrosion product films.
5. There is a need for correlating lab test data with field data (comparison of E_{CORR} and v_{CORR} data from lab and field measurements is not always straightforward).
6. The effects of radiation on corrosion should be considered.
7. Can passive corrosion lead to the formation of an oxide layer that is able to generate mechanical stresses in the concrete?
8. There is a need to obtain realistic mechanical property data of the corrosion products, particularly when assessing the risk of expansive corrosion.
9. There is also a need to understand the real corrosion product layers formed in service—they can be much more complex than in laboratory experiments.
10. There is a need to couple corrosion, radiation and geochemical models.
11. A detailed analysis of the corrosion products films and their electronic properties is required in order to be able to model their behaviour over long time periods, e.g. using the Point Defect Model (PDM) or the high-field growth model.
12. There is a need to carry out tests in solid matrices to allow for mass transport limitations (most of the experiments reported in the literature are performed in solutions that represent cementitious pore water composition). In addition, there may also be different behaviour at the interface between the metal and the surrounding solid matrix. The results from a range of experiments indicate that tests in solution give similar results to those in solid matrices.
13. There is a need for appropriate instrumentation/sensors to determine the ageing of the structures as well as to provide information for the proper design of future facilities.
14. There is a need for a good statistical basis for key corrosion data used in repository performance assessments.
15. There is a need to correlate laboratory test data with data obtained from field measurements, but comparison of corrosion potential and corrosion rate data from laboratory and field measurements is not always straightforward and the results do not always agree.
16. Test duration can play an important role in lifetime predictions of reinforced concrete structures and this should be taken into account in test programmes.

References

ACI: Protection of metals in concrete against corrosion, Committee Report No. ACI 222R–01, American Concrete Institute, Farmington Hills, USA (2001)

Ahmad, Z.: Principles of Corrosion Engineering and Corrosion Control. Elsevier, Amsterdam (2006)

Alexander, M., Otieno, M., Beushausen, H.: Corrosion in cracked concrete—and related issues of structural durability and service life, keynote lecture presented at International Workshop NUCPERF 2012: Long-Term Performance of Cementitious Barriers and Reinforced Concrete in

Nuclear Power Plant and Radioactive Waste Storage and Disposal, RILEM Event TC 226-CNM and EFC Event 351 (2012)
Andrade, C., Zuloaga, P., Martínez, I., Castillo, Á., Briz, S.: Effect of temperature on corrosion parameters and apparent activation energy measured by embedded sensors in pilot container in El Cabril repository. Corros. Eng. Sci. Technol. **46**(2), 182–189 (2011)
Andrade, C., Fullea, J., Toro, L., Martinez, I., Rebolledo, N.: Reinforcement corrosion in alkaline chloride media with reduced oxygen concentrations. In: Proceedings of NUCPERF 2012 Workshop, Long-Term Performance of Cementitious Barriers and Reinforced Concrete in Nuclear Power Plant and Radioactive Waste Storage and Disposal Special issue of The European Physical Journal, vol. 56—2013 (2013)
Angst, U., Elsener, B., Larsen, C.K., Vennesland, Ø.: Critical chloride content in reinforced concrete—a review. Cem. Concr. Res. **39**, 1122–1138 (2009)
Bataillon, C., Bouchon, F., Chainais-Hillairet, C., Desgranges, C., Hoarau, E., Martin, F., Perrin, S., Tupin, M., Talandier, J.: Corrosion modelling of iron based alloy in nuclear waste repository, Electrochim. Acta 55 **15**(1), 4451–4467 (2010)
Bentur, A., Diamond, S., Berke, N.S.: Steel Corrosion in Concrete: Fundamentals and Civil Engineering Practice. E & FN Spon, London (1997)
Berkeley, K.G.C., Pathmanaban, S.: Cathodic Protection of Reinforcement Steel in Concrete. Butterworths & Co, London (1990)
Bertolini, L., Pedeferri, P., Pastore, T.: Stainless steel in reinforced concrete structures. In: Gjørv, O.E., Sakai, K., Banthai, N. (eds.) Concrete under severe conditions 2—environment and loading (CONSEC'98), vol. 1, pp. 94–103. E&FN Spon, London (1998)
Bertolini, L., Elsener, B., Pedeferri, P., Polder, R.: Corrosion of Steel in Concrete: Prevention, Diagnosis, Repair. Wiley-VCH Verlag GmbH & Co., Weinheim (2004)
Brem, M., Büchler, M., Mailand, I.: Online corrosion monitoring and cathodic protection of a steel structure in concrete. In: L'Hostis, V., Gens, R., Gallé, C. (eds.) Proceedings of the RILEM Workshop on Long-Term Performance of Cementitious Barriers and Reinforced Concrete in Nuclear Power Plants and Waste Management, EFC Event 317, RILEM Proceedings PRO 64, pp. 335–342, ISBN: 978-2-31158-072-1 (2009)
Broomfield, J.P.: Corrosion of Steel in Concrete: Understanding, Investigation and Repair, 2nd edn. Taylor & Francis, London (2007)
Caré, S., Berthaud, Y., Ragueneau, F., L'Hostis, V., Millard, A.: Damage of the steel-concrete interface due to accelerated corrosion. In: L'Hostis, V., Gens, R., Gallé, C. (eds.) Proceedings of the RILEM Workshop on Long-Term Performance of Cementitious Barriers and Reinforced Concrete in Nuclear Power Plants and Waste Management, EFC Event 317, RILEM Proceedings PRO 64, pp. 61–70, ISBN: 978-2-31158-072-1 (2009)
Charles, S., Poyet, S., L'Hostis, V.: Impact of moisture parameters on the chloride induced corrosion in concrete. In: L'Hostis, V., Gens, R., Gallé, C. (eds.) Proceedings of the RILEM Workshop on Long-Term Performance of Cementitious Barriers and Reinforced Concrete in Nuclear Power Plants and Waste Management, EFC Event 317, RILEM Proceedings PRO 64, pp. 79–88, ISBN: 978-2-31158-072-1 (2009)
Chitty, W.-J., Berger, P., Dillmann, P., L'Hostis, V., Beranger, G.: Contribution of archaeological analogues to the comprehension of long-term corrosion of concrete reinforcements. J. Phys. IV **136**, 295–304 (2006)
Clear, K.C.: Effectiveness of epoxy-coated reinforcing steel – interim report. Canadian Strategic Highway Research Program (C-SHRP), Ottawa (1992)
Dang, V.H., François, R., L'Hostis, V.: Effects of pre-cracks on both initiation and propagation of re-bar corrosion in pure carbon dioxide. In: Proceedings of NUCPERF 2012 Workshop, Long-Term Performance of Cementitious Barriers and Reinforced Concrete in Nuclear Power Plant and Radioactive Waste Storage and Disposal Special issue of The European Physical Journal, vol. 56—2013 (2013)
Duffó, G.S., Farina, S.B., Arva, E.A., Giordano, C.M., Lafont, C.J.: Studies concerning the durability of concrete vaults for intermediate level radioactive waste disposal: electrochemical monitoring and corrosion aspects. J. Phys. IV **136**, 3–11 (2006)

Duffó, G., Arva, E.A., Schulz, F.M., Vazquez, D.R.: Corrosion behaviour of steel rebars embedded in a concrete designed for the construction of an intermediate-level radioactive waste disposal facility. In: Proceedings of NUCPERF 2012 Workshop, Long-Term Performance of Cementitious Barriers and Reinforced Concrete in Nuclear Power Plant and Radioactive Waste Storage and Disposal Special issue of The European Physical Journal, vol. 56—2013 (2013)

Farina, S.: Corrosion susceptibility of steel drums to be used as containers for intermediate level nuclear waste. In: Proceedings of NUCPERF 2012 Workshop, Long-Term Performance of Cementitious Barriers and Reinforced Concrete in Nuclear Power Plant and Radioactive Waste Storage and Disposal Special issue of The European Physical Journal, vol. 56—2013 (2013)

Farina, S.B., Schulz, F.M., Duffó, G.S.: Corrosion of steel drums containing cemented ion exchange-resins. In: Proceedings of the RILEM Workshop on Long-Term Performance of Cementitious Barriers and Reinforced Concrete in Nuclear Power Plants and Waste Management, EFC Event 317, RILEM Proceedings PRO 64, pp. 53–60, ISBN: 978-2-31158-072-1 (2009)

Fennell, P.: The long-term prediction of corrosion damage in nuclear waste containers. In: Proceedings of the RILEM Workshop on Long-Term Performance of Cementitious Barriers and Reinforced Concrete in Nuclear Power Plants and Waste Management, EFC Event 317, RILEM Proceedings PRO 64, pp. 245–253, ISBN: 978-2-31158-072-1 (2009)

François, R., Castel, A., Vidal, T., Vu, N.-A.: Long-term corrosion behaviour of reinforced concrete structures in chloride environment. J. Phys. IV **136**, 285–293 (2006)

François, R., Castel, A., Zhang, R.: Effect of corrosion on load-bearing capacity and flexural stiffness of reinforced concrete elements. In: Proceedings of the RILEM Workshop on Long-Term Performance of Cementitious Barriers and Reinforced Concrete in Nuclear Power Plants and Waste Management, EFC Event 317, RILEM Proceedings PRO 64, pp. 23–31, ISBN: 978-2-31158-072-1 (2009)

Gens, R., Bel, J., Pourbaix, A., Hélie, M., Wickham, S., Bennett, D.: Corrosion processes and the expected evolution of the BSC–1 Supercontainer design for disposal of Belgian HLW and spent fuel. J. Phys. IV **136**, 13–23 (2006)

Hansson, C.M., Poursaee, A., Jaffer, S.J.: Corrosion of reinforcing bars in concrete. R&D Serial No. 3013, Portland Cement Association (PCA), Skokie, USA (2007)

Harnisch, J., Warkus, J., Raupach, M.: Modelling of reinforcement corrosion: macro cells and time dependence. In: Proceedings of the RILEM Workshop on Long-Term Performance of Cementitious Barriers and Reinforced Concrete in Nuclear Power Plants and Waste Management, EFC Event 317, RILEM Proceedings PRO 64, pp. 71–78, ISBN: 978-2-31158-072-1 (2009)

Hurley, M.F., Scully, J.R.: Chloride threshold levels in clad 316L and solid 316LN stainless steel rebar. Paper presented at CORROSION 2002, Denver, 7–11 April 2002, Paper 02224 (2002)

Jung, H., Kwon, K.-J., Lee, E., Kim, D.-G., Kim, G.Y.: Effect of dissolved oxygen on corrosion properties of reinforcing steel. Corr. Eng. Sci. Technol. **46**(2), 195–198 (2011)

King, F., Padovani, C.: Review of the corrosion performance of selected canister materials for disposal of UK HLW and/or spent fuel. Corros. Eng. Sci. Technol. **46**(2), 82–90 (2011)

Kursten, B., Druyts, F., Macdonald, D.D., Smart, N.R., Gens, R., Wang, L., Weetjens, E., Govaerts, J.: Review of corrosion studies of metallic barrier in geological disposal conditions with respect to Belgian Supercontainer concept. Corros. Eng. Sci. Technol. **46**(2), 91–97 (2011)

Kurtis, K.E., Mehta, K.A.: Critical review of deterioration of concrete due to corrosion of reinforcing steel. In: Malhotra, V.M. (ed.) Proceedings of the fourth CANMET/ACI international conference on durability of concrete, American Concrete Institute publication SP–170, pp. 535–554 (1997)

L'Hostis, V., Amblard, E., Blanc, C., Miserque, F., Paris, C., Bellot-Gurlet, L.: Passive corrosion of steel in concrete in context of nuclear waste disposal. Corros. Eng. Sci. Technol. **46**(2), 177–181 (2011)

Legat, A., Česen, A., Gartner, N., Kuhar, V., Kosec, T.: Use of stainless steel reinforcement in the engineered barriers of LILW repository. In: Proceedings of the RILEM Workshop on Long-Term Performance of Cementitious Barriers and Reinforced Concrete in Nuclear Power Plants and Waste Management, EFC Event 317, RILEM Proceedings PRO 64, pp. 295–303, ISBN: 978-2-31158-072-1 (2009)

Macdonald, D.D., Azizi, O., Saleh, A.: The passivity and corrosion resistance of carbon steel in simulated Boom Clay HLNW environments. In: Proceedings of the RILEM Workshop on Long-Term Performance of Cementitious Barriers and Reinforced Concrete in Nuclear Power Plants and Waste Management, EFC Event 317, RILEM Proceedings PRO 64, pp. 35–44, ISBN: 978-2-31158-072-1 (2009)

Macdonald, D.D., Urquidi-Macdonald, M., Engelhardt, G.R., Azizi, O., Saleh, A., Almazooqi, A., Rosas-Camacho, O Some important issues in electrochemistry of carbon steel in simulated concrete pore water: part 1—theoretical issues. Corros. Eng. Sci. Technol. **46**(2), 98–103 (2011a)

Macdonald, D.D., Saleh, A., Lee, S.K., Azizi, O., Rosas-Camacho, O., Al-Marzooqi, A., Taylor, M.: The electrochemistry of carbon steel in simulated concrete pore water in boom clay repository environments. In: Proceedings of NUCPERF 2012 Workshop, Long-Term Performance of Cementitious Barriers and Reinforced Concrete in Nuclear Power Plant and Radioactive Waste Storage and Disposal Special issue of The European Physical Journal, vol. 56—2013 (2011b)

Mehta, P.K., Monteiro, P.J.M.: Concrete: Structure, Properties and Materials. Prentice-Hall, New Jersey (1993)

Mendoza, A.R.: Corrosion of reinforcing steel in loaded cracked concretes exposed to de-icing salts. Ph.D. thesis, University of Waterloo pp. 16–17 (2003)

Mindess, S., Young, J.F., Darwin, D.: Concrete, 2nd edn. Prentice-Hall Inc, New Jersey (2003)

Miserque, F., Huet, B., Azou, G., Bendjaballah, D., L'Hostis, V.: X-ray photoelectron spectroscopy and electrochemical studies of mild steel FeE500 passivation in concrete simulated water. J. Phys. IV **136**, 89–97 (2006)

Neff, D., Dillmann, P., Burger, E., Demoulin, A., L'Hostis, V.: Thick corrosion layers in reinforced concrete: characterization and properties—a review. In: Proceedings of the RILEM Workshop on Long-Term Performance of Cementitious Barriers and Reinforced Concrete in Nuclear Power Plants and Waste Management, EFC Event 317, RILEM Proceedings PRO 64, pp. 213–219, ISBN: 978-2-31158-072-1 (2009)

Pabalan, R., Chiang, K.-T.: Experimental study and reactive transport modelling of boric acid leaching of concrete. In: Proceedings of NUCPERF 2012 Workshop, Long-Term Performance of Cementitious Barriers and Reinforced Concrete in Nuclear Power Plant and Radioactive Waste Storage and Disposal Special issue of The European Physical Journal, vol. 56—2013 (2013)

Pabalan, R.T., Glasser, F.P., Pickett, D.A., Walter, G.R., Biswas, S., Juckett, M.R., Sabido, L.M., Myers, J.L.: Review of literature and assessment of factors relevant to performance of grouted systems for radioactive waste disposal. Report No. CNWRA 2009–001, Center for Nuclear Waste Regulatory Analyses (CNWRA), San Antonio, US (2009)

Pabalan, R.T., Yang, L., Chiang, K.-T.: Boric acid corrosion of concrete rebar. In: Proceedings of NUCPERF 2012 Workshop, Long-Term Performance of Cementitious Barriers and Reinforced Concrete in Nuclear Power Plant and Radioactive Waste Storage and Disposal Special issue of The European Physical Journal, vol. 56—2013 (2013)

Pourbaix, A., L'Hostis, V.: Passivation, localised corrosion and general corrosion of steel in concrete and bentonite: theory and experimentals. J. Phys. IV **136**, 71–78 (2006)

Saheb, M., Marsal, F., Matthiesen, H., Neff, D., Dillmann, P., Pellegrini, D.: Fluctuations of redox conditions in radioactive waste disposal cell: characterisation of corrosion layers formed on archaeological analogues. Corros. Eng. Sci. Technol. **46**(2), 199–204 (2011)

Saleh, A., Azizi, O., Rosas-Camacho, O., Al-Marzooqi, A., Macdonald, D.D.: Some important issues in electrochemistry of carbon steel in simulated concrete pore water: Part 2—Experimental. Corros. Eng. Sci. Technol. **46**(2), 104–110 (2011)

Schulz, F.M., Arva, E.A., Duffó, G.S.: Studies concerning the reinforcement corrosion of concrete vaults for intermediate-level radioactive waste disposal. In: L'Hostis, V., Gens, R., Gallé, C. (eds.) Proceedings of the RILEM Workshop on Long-Term Performance of Cementitious Barriers and Reinforced Concrete in Nuclear Power Plants and Waste Management, EFC Event 317, RILEM Proceedings PRO 64, pp. 13–21, ISBN: 978-2-31158-072-1 (2009)

Smart, N.R., Rance, A.P., Winsley, R.J., Fennell, P.A.H., Reddy, B., Kursten, B.: The effect of irradiation on the corrosion of carbon steel in alkaline media. In: Proceedings of the RILEM Workshop on Long-Term Performance of Cementitious Barriers and Reinforced Concrete in Nuclear Power Plants and Waste Management, EFC Event 317, RILEM Proceedings PRO 64, pp. 45–52, ISBN: 978-2-31158-072-1 (2009)

Smart, N.R., Fennell, P.A.H., Rance, A.P., Winsley, R.J., Reddy, B., Kursten, B.: Experimental studies of the effect of irradiation on the anaerobic corrosion of carbon steel in relation to the Belgian supercontainer concept. EPJ Web Conf. **12**, 02003 (2011)

Smart, N.R., Rance, A.P., Fennell, P.A.H., Kursten, B.: The anaerobic corrosion of carbon steel in alkaline media—phase 2 results. In: Proceedings of NUCPERF 2012 Workshop, Long-Term Performance of Cementitious Barriers and Reinforced Concrete in Nuclear Power Plant and Radioactive Waste Storage and Disposal Special issue of The European Physical Journal, vol. 56—2013 (2013)

Tikalsky, P.J., Beh, D.E.: Synthesis guide to best practices for corrosion resistant concrete. Utah Department of Transportation, Salt Lake City, Utah, USA, report UT-08.27 (2008)

Wegemar, B.: Corrosion failure of a BWR embedded reactor containment liner. J. Phys. IV **136**, 257–262 (2006)

Weiss, C.A., Morefield, S.W., Malone, P.G., Koenigstein, M.L.: Role of reactive vitreous coatings in reinforced concrete in nuclear plant construction. In: Proceedings of the RILEM Workshop on Long-Term Performance of Cementitious Barriers and Reinforced Concrete in Nuclear Power Plants and Waste Management, EFC Event 317, RILEM Proceedings PRO 64, pp. 287–294, ISBN: 978-2-31158-072-1 (2009)

Winsley, R.J., Smart, N.R., Rance, A.P., Fennell, P.A.H., Reddy, B., Kursten, B.: Further studies on the effect of irradiation on the corrosion of carbon steel in alkaline media. Corros. Eng. Sci. Technol. **46**(2), 111–116 (2011)

Chapter 4
Ageing Management of Concrete Structures of NPPs

Etienne Gallittre, James Mok and Jean Luc Valfort

Abstract The subject of ageing management of concrete structures in Nuclear Power Plants (NPPs) has been discussed extensively in the past. Recently this subject has become more important because many of the existing Nuclear Power Plants in the world are reaching or near their original design life. Rilem Technical Committee TC-226 organised a series of conferences and workshops in the past to gather the information and experiences from the international communities in order to collect the current experience, and to identify the knowledge gaps and future challenges in ageing management of concrete structures of NPPs. A summary of this information is presented in this chapter.

Keywords Plan-do-check-act cycle · Optimisation · Operation · Inspection · Monitoring · Maintenance

4.1 Introduction

Concrete can be a very durable construction material under favourable environmental conditions and its performance as concrete containment and other safety-related structures in nuclear power plants has been good. Experience shows that ageing degradation can be a result of exposure to aggressive environments, excessive structural loads, accident conditions, use of unsuitable materials, poor material and construction quality, and the lack of or inadequate maintenance or inspections, etc. (Naus 2010; IAEA 1998, 2009; USNRC 1997). Furthermore in some cases, materials or components of the concrete structures may need to be repaired or upgraded to meet modern codes requirements for environments and loadings conditions, and/or due to unavailability of original specified materials.

E. Gallittre · J.L. Valfort
EdF, Villeurbanne, France

J. Mok (✉)
CNSC, Ottawa, ON, Canada
e-mail: James.Mok@cnsc-ccsn.gc.ca

V. L'Hostis and R. Gens (eds.), *Performance Assessment of Concrete Structures and Engineered Barriers for Nuclear Applications*,
RILEM State-of-the-Art Reports 21, DOI 10.1007/978-94-024-0904-8_4

However, the functional or performance requirements defined by the design basis and safety analysis of the concrete structures should not be affected provided the structures are kept in good maintenance by timely detection and mitigation of degradations with an ageing management program (AMP).

4.2 Ageing Management of Concrete Structures of NPPs

An AMP is a set of integrated engineering, operation and maintenance actions to control ageing degradation of systems, structures and components (SSCs) within acceptable limits. Ageing management of concrete structures of NPPs is achieved by periodic inspection of the accessible areas and other techniques are used for monitoring of ageing of the areas that are not readily accessible. An AMP for concrete structures should clearly identify the effective actions and measures for managing ageing in a timely manner. It should provide performance indicators which measures the effectiveness of the current practices based on the evaluations of ageing and condition assessments of the concrete structures.

The AMP can consist of existing plants programs such as periodic inspections, maintenance, condition assessments, obsolescence management and system health monitoring, etc. The AMP of concrete structures can be implemented under the operators' Quality Management System for the NPPs. The AMP should provide a systematic and integrated framework for coordinating programs and activities relating to the understanding, control, monitoring and mitigation of ageing of concrete structures which is based on the Deming's PLAN-DO-CHECK-ACT cycle of ageing management as illustrated in Fig. 4.1 (IAEA 2009).

The summary on the topic of ageing management from the series of conferences and workshops organised by Rilem Technical Committee TC-226 is presented below. It is divided according to the four elements of the Deming's PLAN-DO-CHECK-ACT cycle ageing management above.

4.2.1 "PLAN"—Development and Optimisation of Activities for Ageing Management

The "PLAN" activity in Fig. 4.1 refers to the coordination, integration, maintenance and improvement activities for ageing management of SSCs. This should include but not limited to documenting the regulatory requirements, safety criteria and relevant activities, describing the coordination mechanism and processes, and improving effectiveness of ageing management based on current understanding, self-assessment and peer review. The implementation of the AMP should also include an evaluation of effectiveness of the integrated AMP.

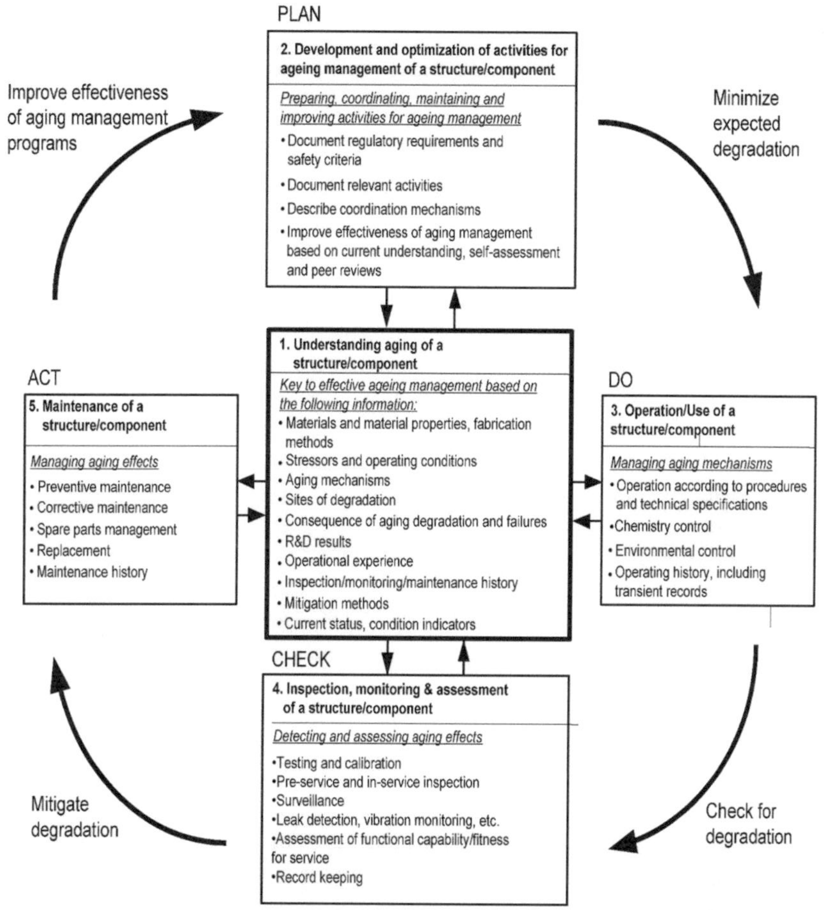

Fig. 4.1 Systematic approach to managing ageing of a structure or component. *Source* IAEA Safety Standards "Ageing Management for Nuclear Power Plants", Safety Guide No. NS-G-2.12

Lipar (2010) presented the IAEA activities on Plant Life Management for Safe Long-Term Operation. IAEA Safety Standards provide a set of guidelines and recommendations for the management of ageing of SSCs important to safety in NPPs. These sets of guidelines and recommendations mainly focused on physical ageing but they also include management of obsolescence, operating experience, quality management, etc. The need for an ageing management program for the NPPs is reinforced by the fact that many of the current NPPs are undergoing refurbishment and extended long-term operation. An AMP should follow the nine generic attributes for effective ageing management. IAEA has also recommended that an International Generic Ageing Lessons Learned (IGALL) document supporting a systematic approach to managing ageing of SSCs for safer operation of existing and future NPPS.

Mok et al. (2010) presented the Canadian Nuclear Regulator's perspective for the ageing management of concrete containment structures (CCSs) of NPPs in Canada. It was noted that the operators of NPPs should have appropriate programs and activities in place to manage ageing effects and obsolescence issues for CCSs. A number of existing programs and activities at the NPPs can contribute to an effective integrated AMP for the CCSs and these can include periodic inspection and testing, preventive and corrective maintenance, equipment reliability and qualification, condition assessment, system health monitoring, procurement, spare parts and obsolescence management as well as research and development, etc. The effectiveness of the overall integrated AMP should be periodically reviewed using feedback from the operating performance, inspection and maintenance histories, event reports, information from the results of research and development, self assessments, and generic operating experience, etc.

Gallittre and Dauffer (2010) presented a specific procedure based on IAEA technical documents and guidelines for the ageing management of French NPP concrete structures for a 60-year service life. The ageing phenomenon is treated from a safety point of view leading to a documented discussion on every ageing mechanism considered. The procedure is based on ageing data sheets especially for critical degradation, mechanisms and the rate of degradation. The data sheets developed are analysed through an ageing management process. Detailed reports are written for important safety related components such as containment buildings (post-tensioned walls and pre-stress losses, steel liner) or for specific ageing mechanisms such as alkali-silica reaction (ASR) or delayed ettringite formation (DEF). The reports are periodically reviewed every 5 years.

Tcherner et al. (2010) presented some site-specific AMPs for nuclear reactors and waste management facilities' civil structures in Canada. The paper describes some selected components of an AMP to facilitate effective implementation of the program and optimum use of resources. Lessons learned in the process of implementation of the site-specific AMPs are also discussed. There is a need for the development of NDE techniques for detecting defects in reinforced concrete structures for continued operation. It was noted that the degree of deterioration mostly depends on the aggressiveness of the environment and initial quality of the structure as constructed. In addition to degradation of joint sealants and non-metallic liners and coatings, exposure to elevated temperatures/temperature fluctuations and leaching of calcium from the concrete facilitated by exposure to flowing demineralised or acidic water for concrete and corrosion for steel components were noted to be the most typical ageing stressors for nuclear civil structures.

Vesikari et al. (2010) presented a service life management system for concrete structures in Finnish NPPs. The system is predictive, probabilistic and life-cycle based and takes into account several aspects of structural lifetime quality, such as performance, condition, financial costs and environmental impacts. It contains an automatic condition guarding system which is able to trigger maintenance, repair and rehabilitation actions when the predicted limit state of condition is exceeded with a maximum allowable probability. Through the service life management system, the accepted structural performance and uninterrupted service of concrete structures are ensured during the planned operating lifetime of the Finnish NPPs.

Al-Neshawy et al. (2012) presented a paper on the continuation of the development of a computerised ageing management system for the concrete structures in the Finnish NPPs. The computerised ageing management system provides access to the structural, material and environmental information and to various design application with appropriate methodologies and optimisation processes, which are required for the designers and maintainers of a nuclear power plants. The central database is developed for assembling and systematically organising the information gathered from structural design, maintenance activities, repairing of nuclear power plants structures, operating and environmental condition, in-service performance and other type of data.

Tangestanian et al. (2010) presented their experience in the implementation of the life cycle management program and condition assessment of major components. It was noted that through interviews with system responsible engineers (SREs) that a systematic approach to training for ageing management, equipment reliability, criticality ranking, licensing basis, examples of degradation mechanisms, operating experience, roles and responsibilities of SREs, etc., is required. Consideration should also be given to the use of third-party peer reviews to obtain an independent assessment to establish whether the life cycle management plan is consistent with generally accepted practices and to identify areas for improvement. The operator of the NPP should update the life cycle management plan and interfacing programs, and their implementation, to improve their effectiveness based on the results of the review as appropriate.

4.2.2 "DO"—Operation and Use of Structures and Components

In Fig. 4.1, the "DO" activity involves the management of ageing mechanisms by minimising the expected degradation of structures or components through their operation or use in accordance with the approved operating procedures and technical specifications. This also includes maintain operating conditions within design limits, establish proper chemistry control and environmental control, and maintaining operating history including transient records.

Weiss et al. (2009) presented a paper on the role of innovative reactive vitreous coatings in reinforced concrete in nuclear plant construction. Reinforced concrete is the basic construction material in NPPs in the United States. Concrete condition surveys conducted at NPPs have shown that problems can develop during operations that are typical of ageing reinforced concrete structural members. Corrosion of the embedded steel is especially a problem where the concrete is exposed to moisture and oxygen can diffuse down to the concrete-steel interface and cause rust formation, expansion and cracking. The recent development of new engineered coatings for reinforcing steel that are based on using an alkali-resistant vitreous enamel to coat the steel and to fuse reactive calcium silicates and aluminates to the surface of the steel can be used to address both the bonding and the corrosion problem in reinforced concrete.

Mozaryn and Kokowska (2009) presented a paper on the service life of coating systems applied on cooling towers. It was noted that a visual inspection of the coating systems for 19 repaired and protected cooling towers were carried out. The coating systems were applied as protection on both the internal and external sides of the cooling towers concrete shells. A visual assessment of the coating systems' appearance and their adhesion to the concrete surfaces were made on site. Water permeability, water vapour permeability, carbon dioxide permeability and coating thickness measurements, were all tested on samples taken from the Cooling Towers, in the laboratory at the Building Research Institute in Warsaw. The assessment of any change in the technical performance and properties of the coating systems after more than 12 years in service was also made by comparison with freshly applied materials. The testing of the coatings systems confirmed their protective efficiency and ability to make a positive contribution to the durability of reinforced concrete Cooling Towers and similarly exposed structures.

Oxfall et al. (2012) presented a paper on the moisture levels and the drying potential of the concrete inside Swedish reactor containments. Due to high temperature during operation, the concrete within the containment may act as a moisture source contributing to the humidity. The moisture content of the concrete may affect the environment conditions inside the containments and hence the moisture distribution should be determined. A measurement setup was developed and tested. Preliminary results showed that the concrete was still drying after 30–35 years of exposure inside the containment.

4.2.3 "CHECK"—Inspection, Monitoring and Assessment of Structures and Components

The "CHECK" activity in Fig. 4.1 of a systematic and integrated approach to ageing management includes the timely inspection, monitoring and assessment of SSCs. Data identified in the AMP should be collected and recorded to provide a basis for decisions on the type and timing of ageing management actions. Feedback from the utilities on the methods of detection of defects and their repair and or mitigation would be beneficial and provide continuous improvement to the AMP.

Visual inspection is the most common type of non-destructive examinations (NDE) to detect ageing effects of concrete structures. Visual inspection of accessible surfaces of the concrete structures are expected to detect and define areas of ageing-related distress that result in visible effects on the surface of the structures, e.g. cracking, spalling, volume change, cement-aggregate separation, mechanical degradation, or moisture movement. Visual inspection can be supplemented by other NDE methods such as rebound hammer, radar, audio, infrared thermography, ultrasonic pulse velocity, tomography, leakage rate test, and instrumentation, etc.

Philipose and Frank (2009) presented a paper on the ageing management of CANDU concrete containment buildings (CCB). The main safety function of the CCB is the leak tightness that it needs to maintain under a design pressure accident. The leak tightness of the CCB can be assured by carrying out the periodic in-service leakage rate test and its leak rates under test pressures must be maintained below acceptable limits. The concrete strain measurements collected in the last 30 years during leak rate tests indicated that the CCB is functioning in satisfactory manner and the ageing effects are not significant. It was also suggested that although the instruments were installed for the purpose of validation of design assumptions and the integrity of the CCB, they can also be used as an ageing management tool for the future years.

Reichling et al. (2009) presented a paper on a robotic system for simultaneous diagnoses of reinforced concrete structures. In the current industry approaches, measurement of various degradation parameters are feasible but often only one parameter is measured at a time. In order to accelerate the measurement of various degradation parameters of concrete structures, a project was initiated in Germany to develop a robotic system which is able to move over large horisontal areas of reinforced concrete structures and carry out non-destructive measurements automatically with various testing instruments integrated. The condition of the concrete structure is assessed based on the investigation of the whole concrete surface by recording useful key parameters such as electrochemical potentials, concrete cover, carbonation depth, chloride profiles, cracks, delaminations and condition of the reinforcement, etc. By combining the different methods within only one measurement operation, an analysis can be accomplished faster and more accurate. Furthermore, detailed maps are created to locate all critical areas. By means of periodical measurements the life-time prognoses of structures can continuously be updated.

Schneck (2009) presented a paper on the informative results from a non-destructive corrosion survey of the soffit of a post-tensioned bridge deck. Surface resistivity and rest potential on the deck surface and on the soffit of the bridge deck were measured. It is shown that non-destructive, full scale corrosion surveys are possible without traffic disruptions and the locations of possible damage can be inspected precisely for safe and object related data evaluation. It is indicated that non-destructive, full scale corrosion surveys are possible without traffic disruptions. The extent and appropriate timing of repairs are determined and planning can be concentrated on necessary areas. This can result in cost efficiency and significantly improved level of durability of repairs.

Brem et al. (2009) presented a paper on online corrosion monitoring and cathodic protection of a steel structure in concrete. In this case, corrosion of a steel structure clamped to a reinforced concrete structure was considered. The joint between the concrete and the steel structure was filled with rock wool. During the first years of operation, water mixed with boric acid contaminated the concrete at the area of the joint. This resulted in an acid attack causing corrosion at the interface where the steel was in contact with the concrete. Potential measurements allowed a detailed assessment of the corrosion of the structure. It was found that the relative humidity in the joint decreased and the concrete resistivity increased (drying). With the monitoring system in place, it is possible to observe the reduction in corrosion rate. The present case shows that it is essential to understand the corrosion mechanism prior to taking measures to improve the structural integrity.

Coppel et al. (2012) presented the approach developed by EDF with respect to the apprehension of risks of internal expansion of the concrete on nuclear structures of NPPs. It was noted that the internal expansion of concrete can be caused by alkali aggregate reaction (AAR) and/or delayed ettringite formation (DEF). These pathologies can cause internal expansion of concrete, cracking and changes in mechanical properties. The prevention of these phenomena has not historically been subject to special requirements in the design and construction of operating French Nuclear Power Plants. In France, recommendations for preventing these phenomena date from 1994 for the AAR and 2007 for the EDF. These ageing phenomena have been the subject of a lot of research over the past few years and progress has been made in the prevision and management of structures subjected to internal expansion of concrete.

Gallittre et al. (2012) presented a paper on the recent activities of an OEAD/NEA working group on the investigation of the comparative advantages and disadvantages of various post-tensioning techniques in reactor containment. This work deals mainly with pre-stressed grouted cables modelling, with PCCV monitoring and also with other items directly linked to cables post-tensioning technologies. Depending on each country's practice, the cables of PCCV can be greased or cement grouted. It was noted that greased cables has the main advantage of allowing direct in-service measurement of the pre-stress force in the cables. For the grouted technology, it has to be very well managed during the construction phases in order to have adequate corrosion protection from the permanent cement grout around the cables. During the plant operation, global strain monitoring system for full pressure test survey and for pre-stress losses assessments will likely be required. This new work is in progress and the final result is expected to be available before end of year 2014.

Wiggenhauser and Niederleithinger (2012) presented innovative ultrasonic techniques for the inspection and monitoring of large concrete structures. Recently developed multi-offset arrays of ultrasonic transducers facilitate not only the deployment of many sensors to the surface at one time but also the application of advanced data processing schemes. The system consists of ten rows of four transducers. The transducers of each row are connected and work together to avoid noisy results. All combinations are switched electronically by the system software. All data are processed by the system using the established SAFT technique. The imaging result is displayed almost instantaneously on the screen. This allows on-site interpretation and corrections in the measurement setup. Larger arrays to allow the investigation of very thick structures are under development. Ultrasonic techniques to image structures and to monitor changes inside large concrete structures have been improved widely in the last decade. These improvements include sensors and sensor arrangement, array techniques, automated data acquisition as well as various data processing techniques. Some of the new methods may require further research and development. However, a set of tools is available for practical application, which will help in the assessment of ageing large concrete structures.

4.2.4 "ACT"—Maintenance of Structures and Components

The "ACT" activity in Fig. 4.1 refers to the timely mitigation and correction of component degradation by the implementation of maintenance, repair and replacement activities. It includes preventive maintenance, corrective maintenance, spare parts management, and replacement, etc.

Aldea et al. (2010) presented an AMP for the reactor building at a Canadian NPP with the specific conditions, administration and operation protocols to deal with the ageing degradations at the station. This includes the development of a specific inspection, monitoring and maintenance/repair plan and condition assessment of the concrete containment structure. Repair methodologies of the deficiencies identified were recommended to ensure the ability of the reactor building to perform its design function and the safe and reliable operation of the plant.

Cremona et al. (2010) reported on the extensive and focused research activities conducted under the APPLET project to conduct research to understand concrete ageing degradation and to assess residual life time of reinforced concrete structures by using a probabilistic and predictive method. The material testing and research on new concrete laboratory specimen and concrete samples collected from new and existing concrete structures) were focused on concrete-environment interfaces, electro-chemical diagnosis methods, mechanical behaviour of degraded structures and variability and measurement uncertainty to support probabilistic assessment. The data obtained from the APPLET project would serve as a large resource base for further research in concrete degradation and residual life assessment of concrete structures. The numerical models which have been developed and implemented in various programs will constitute a strong platform for further developments.

Cho et al. (2010) recently developed a new Structural Life Management System (SLMS) to respond to the needs that emerged during the decade following the development and operation of the first SLMS for Korean NPPs in 1998. The new SLMS is featured with centralised control of the security of the data and the improved maintenance efficiency through systematic digitalisation of synthetic history records and items of the structure necessary to its maintenance. In addition, field monitoring data related to carbonation and chloride attack provides opportunity to manage long-term durability of the structures. The capability to compute the integrity indices of the structures instantaneously with the input of degradation data or recent inspection records using an integrity assessment program.

Cornish-Bowden et al. (2010) presented their study on the operating experience on surveillance and maintenance of buried pipes in several industry applications. In the oil and gas industry, pipelines are generally well known and located. On the other hand, in the nuclear industry buried piping degradation has become a major issue since the last few years because assessment of the structures has been post-poned and neglected for a significant period of time. The research work done for the development of instrumented vehicles suitable for nuclear buried pipes, and the change of policy of maintenance in the water supply industry should help the operators to improve their buried pipes safety and performance in the NPPs.

4.3 Conclusion

Significant attention was given to the development and implementation of AMP in NPPs recently, in particular for life extension projects because of the need to demonstrate the SSCs of NPPs can perform as designed for the extended service life. Based on the information collected from the various past RILEM conferences and workshops such as NUCPERF 2009, AMP 2010 and NUCPERF 2012, the approach to develop an AMP for concrete structures is generally well documented and can be found in many sources such as IAEA and USNRC.

AMP should be developed and implemented by the operators of NPPs and it can consist of existing plants programs such as periodic inspections, maintenance, condition assessments, obsolescence management and system health monitoring, etc. However, it should include all the attributes of an effective and integrated AMP as described in IAEA NS-G-2.12. Continuous improvement of the AMP should also be performed based on feedback of operating experience, results from research and development, self-assessment and peer reviews.

Inspection techniques and testing methods for NDE of concrete structures continue to evolve. There are on-going researches and studies of new inspection techniques and testing methods for NDE being conducted by various stakeholders. Operators of NPP should obtain feedback from the results of these researches and studies, and evaluate the effectiveness of their current inspection techniques and testing methods in the plant. Knowledge of inspection personnel in these areas should be kept up to date and trained with the latest knowledge available.

4.4 RILEM TC-226-CNM Recommendations for Future R&D

There are remaining challenges in the areas of ageing management of concrete structures in NPP and NDE. These include but not limited to:

- Inspection techniques for grouted tendons in pre-stressed systems
- Modelling techniques for non-grouted tendons in pre-stressed systems
- Effective ageing management for concrete structures with AAR and DEF
- Inspection of inaccessible areas such as emergency cooling water tank, etc.

Currently there are a number of international initiatives regarding ageing management and NDE for concrete structures of NPPs. IAEA initiated a programme "International Generic Lessons Learned" (I-GALL) in 2009. The programme is to assist regulators and operators of nuclear power plants to maintain the required safety level during operation taking into account the existing degradation mechanisms and to provide generic support in identification of criteria and practices at the national level applicable to continued operation of the plants.

Within the framework of OECD/NEA, two working groups are currently in progress. One of the working groups is to investigate the post-tensioning methodologies for concrete containment structures with the focus on the development of some inspection guidelines for grouted or non-grouted tendons. The other working group is on the non-destructive evaluation of thick walled concrete structures. Other topics such as the effect and ageing management of alkaline-aggregate reaction will need more in-depth investigation. Another field of progress from the industry is a better use of the data coming from surveillance and monitoring using more reliable instrumentation.

References

Aldea, C.M., Shenton, B., DeMerchant, M.M., Gendron T.: Ageing management program of the reactor building concrete at Point Lepreau Generating Station, AMP 2010. In: International Workshop on Ageing Management of Nuclear Power Plants and Waste Disposal Structures (EFC Event 334), The European Physical Journal, vol. 12—2011 (2010)

Al-Neshawy, F., Piironen, J., Sistonen, E., Vesikari, E., Tuomisto, M., Hradil, P., Feirrera, M.: Developing a computerised ageing management system for concrete structures in finnish nuclear power plants. In: Presentation made at the NUCPERF 2012 Workshop, Long-Term Performance of Cementitious Barriers and Reinforced Concrete in Nuclear Power Plant and Radioactive Waste Storage and Disposal (2012)

Brem, M., Büchler, M., Mailand, I.: Online corrosion monitoring and cathodic protection of a steel structure in concrete. In: L'Hostis, V., Gens, R., Gallé, C. (eds.) Proceedings of the RILEM Workshop on Long-Term Performance of Cementitious Barriers and Reinforced Concrete in Nuclear Power Plants and Waste Management, EFC Event 317, RILEM Proceedings PRO 64, ISBN: 978-2-31158-072-1 (2009)

Cho, M., Kwon, B., Lee, C., Lee, S.: Structural life management system of nuclear power plant in Korea. In: Proceedings of the AMP 2010 Workshop, International Workshop on Ageing Management of Nuclear Power Plants and Waste Disposal Structures (EFC Event 334), ISBN 978-2-7272-0230-1 (2010)

Coppel, F., Lion, M., Vincent, C., Roure, T.: Approaches developed by EDF with respect to the apprehension of risks of internal expansion of the concrete on nuclear structures: management of operating power plants and prevention for new power plants. In: Presentation made at the NUCPERF 2012 Workshop, Long-Term Performance of Cementitious Barriers and Reinforced Concrete in Nuclear Power Plant and Radioactive Waste Storage and Disposal Special (2012)

Cornish-Bowden, I., Thillard, G., Capra, B.: Capitalisation of operating experience on surveillance and maintenance of buried pipes in several industrial applications, AMP 2010. In: International Workshop on Ageing Management of Nuclear Power Plants and Waste Disposal Structures (EFC Event 334), The European Physical Journal, vol. 12—2011 (2010)

Cremona, C., Adélaide, L., Berthaud, Y., Bouteiller, V., L'Hostis, V., Poyet, S., Torrenti J.M.: Probabilistic andpredictive performance-based approach for assessing reinforced concrete structures lifetime: the applet project. In: Proceedings of the AMP 2010 Workshop, International Workshop on Ageing Management of Nuclear Power Plants and Waste Disposal Structures (EFC Event 334), The European Physical Journal, vol. 12 http://dx.doi.org/10.1051/epjconf/20111201004 (2011)

Gallittre, E., Dauffer, D.: Ageing management of french NPP civil work structures, AMP 2010. In: International Workshop on Ageing Management of Nuclear Power Plants and Waste Disposal Structures (EFC Event 334), The European Physical Journal, vol. 12—2011 (2010)

Gallittre, E., Varpasuo, P., Huerta, A., Valfort, J-L.: Post-tensioning methodologies for containment building: greased or cement grouted tendons—consequences on monitoring, periodic testing and modelling activities. In: Presentation made at the NUCPERF 2012 Workshop, Long-Term Performance of Cementitious Barriers and Reinforced Concrete in Nuclear Power Plant and Radioactive Waste Storage and Disposal (2012)

IAEA: International Atomic Energy Agency, IAEA Safety Guide NS-G-2.12, Ageing management for nuclear power plants (2009)

IAEA: International Atomic Energy Agency, IAEA TECDOC-1025, Assessment and management of ageing of major nuclear power plant components important to safety: concrete containment buildings (1998)

Lipar, M.: IAEA Activities on plant life management for safe long-term operation, AMP 2010. In: International Workshop on Ageing Management of Nuclear Power Plants and Waste Disposal Structures (EFC Event 334), The European Physical Journal vol. 12—2011 (2010)

Mok, J., Kirkhope, K., Lim, D.: Ageing Management of Concrete Containment Structures in Canada-Regulatory Perspective. In: Proceedings of the AMP 2010 Workshop, International Workshop on Ageing Management of Nuclear Power Plants and Waste Disposal Structures (EFC Event 334), ISBN 978-2-7272-0230-1 (2010)

Mozaryn, T., Kokowska, J.: Service life of coating systems applied on cooling towers—laboratory study and in-situ investigations. In: L'Hostis, V., Gens, R., Gallé, C. (eds.) Proceedings of the RILEM Workshop on Long-Term Performance of Cementitious Barriers and Reinforced Concrete in Nuclear Power Plants and Waste Management, EFC Event 317, RILEM Proceedings PRO 64, ISBN: 978-2-31158-072-1 (2009)

Naus, D.: Overview of activities in the U.S. related to continued service of NPP concrete structures, AMP 2010. In: International Workshop on Ageing Management of Nuclear Power Plants and Waste Disposal Structures (EFC Event 334), The European Physical Journal, vol. 12—2011 (2010)

Oxfall, M., Hassanzadeh, M., Johansson, P.: Moisture levels and the drying potential of the concrete inside Swedish reactor containments. In: Proceedings of NUCPERF 2012 Workshop, Long-Term Performance of Cementitious Barriers and Reinforced Concrete in Nuclear Power Plant and Radioactive Waste Storage and Disposal Special issue of The European Physical Journal, vol. 56—2013 (2012)

Philipose, K., Frank, E.: Ageing management of CANDU concrete containment buildings. In: L'Hostis, V., Gens, R., Gallé, C. (eds.) Proceedings of the RILEM Workshop on Long-Term Performance of Cementitious Barriers and Reinforced Concrete in Nuclear Power Plants and Waste Management, EFC Event 317, RILEM Proceedings PRO 64, ISBN: 978-2-31158-072-1 (2009)

Reichling, K., Raupach, M., Wiggenhauser, H., Stoppel, M., Dobmann, G., Kurz, J.: BETOSCAN—a robotic system for simultaneous diagnoses of reinforced concrete structures. In: L'Hostis, V., Gens, R., Gallé, C. (eds.) Proceedings of the RILEM Workshop on Long-Term Performance of Cementitious Barriers and Reinforced Concrete in Nuclear Power Plants and Waste Management, EFC Event 317, RILEM Proceedings PRO 64, ISBN: 978-2-31158-072-1 (2009)

Schneck, U.: Informative results from a non-destructive corrosion survey of the soffit of an 80 cm thick post-tensioned bridge deck. In: L'Hostis, V., Gens, R., Gallé, C. (eds.) Proceedings of the RILEM Workshop on Long-Term Performance of Cementitious Barriers and Reinforced

Concrete in Nuclear Power Plants and Waste Management, EFC Event 317, RILEM Proceedings PRO 64, ISBN: 978-2-31158-072-1 (2009)
Tangestanian, P., Keshavarz, A., Wiebe, A.: Integrated ageing management program at Ontario Power Generation. In: Proceedings of the AMP 2010 Workshop, International Workshop on Ageing Management of Nuclear Power Plants and Waste Disposal Structures (EFC Event 334), ISBN 978-2-7272-0230-1 (2010)
Tcherner, J., Aziz, T., Philipose, K.: Experience Gained in Ageing Management of Nuclear Civil Structures. In: Proceedings of the AMP 2010 Workshop, International Workshop on Ageing Management of Nuclear Power Plants and Waste Disposal Structures (EFC Event 334), ISBN 978-2-7272-0230-1 (2010)
USNRC NUREG-1611: Ageing management of nuclear power plant containments for license renewal (1997)
Vesikari, E., Hiltunen, V., Mattila, A.: Service life management system of concrete structures in nuclear power plants. In: Proceedings of the AMP 2010 Workshop, International Workshop on Ageing Management of Nuclear Power Plants and Waste Disposal Structures (EFC Event 334), ISBN 978-2-7272-0230-1 (2010)
Weiss, C., Morefied, S.W., Malone, P.G., Koenigstein, M.L.: Role of innovative reactive vitreous coatings in reinforced concrete in nuclear plant construction. In: L'Hostis, V., Gens, R., Gallé, C. (eds.) Proceedings of the RILEM Workshop on Long-Term Performance of Cementitious Barriers and Reinforced Concrete in Nuclear Power Plants and Waste Management, EFC Event 317, RILEM Proceedings PRO 64, ISBN: 978-2-31158-072-1 (2009)
Wiggenhauser, H., Niederleithinger, E.: Ultrasonic techniques for inspection and monitoring of large concrete structures. In: Proceedings of NUCPERF 2012 Workshop, Long-Term Performance of Cementitious Barriers and Reinforced Concrete in Nuclear Power Plant and Radioactive Waste Storage and Disposal Special issue of The European Physical Journal, vol. 56—2013 (2012)

Chapter 5
Waste Management and Performance Assessment

Jacob Philip and Steve Williams

Abstract Cement barriers play an important role in many radioactive waste management and disposal systems where they may provide physical and/or chemical barriers to limit the release of many radionuclides. Therefore, an understanding of how cement barriers are likely to evolve over the very long timescales that may be considered in safety cases and performance assessments is important. This can build confidence in the basis of a performance assessment or may reduce the need to make simple overly-conservative assumptions. This section begins with an overview of the role of cementitious materials in radioactive waste management and disposal, performance assessment and the treatment of uncertainty. It then summarises relevant papers presented at the NUCPERF 2009, NUCPERF 2012 and AMP 2010 workshops on the topics of the role and durability of cement barriers in waste storage and disposal and on the impact of gas generation and carbonation of cement systems. Some recommendations for future R&D are then made after drawing conclusions.

Keywords Disposal systems · Chemical and physical barriers · Radionuclides · Treatment of uncertainties

5.1 Introduction

Cementitious engineered materials have been used or proposed in a variety of waste management systems because these materials can be formulated with desirable performance characteristics (e.g. hydraulic isolation, chemical isolation and structural stability). Cementitious barriers are commonly engineered with a goal of achieving specific performance (e.g. minimisation of hydraulic conductivity,

J. Philip
NRC, Washington, DC, USA

S. Williams (✉)
Radioactive Waste Management, Didcot, UK
e-mail: Steve.Williams@nda.gov.uk

V. L'Hostis and R. Gens (eds.), *Performance Assessment of Concrete Structures and Engineered Barriers for Nuclear Applications*,
RILEM State-of-the-Art Reports 21, DOI 10.1007/978-94-024-0904-8_5

provision of required porosity or diffusivity or the maintenance of suitable chemical conditions). However a simple performance goal may not be optimum when practical considerations of design and performance characteristics are considered simultaneously. Laboratory-scale optimised designs may also have full-scale characteristics that are less than ideal (Esh et al. 2011).

Typical generic safety functions of cements employed in multi-barrier disposal systems may include providing a stable low solubility matrix that limits the release of many radionuclides by dissolving slowly (the wasteform), acting as a partial barrier limiting the access of water to the wasteform (the container), or conditioning the chemical characteristics of groundwater and porewater to reduce container corrosion or limit the dissolution of radionuclide-containing phases (the backfill) (Nuclear Decommissioning Authority 2010). Safety function indicators (a measurable or calculable property of a component that indicates the extent to which the safety function is fulfilled) and safety function indicator criteria (a quantitative limit such that if the safety function indicator to which it relates fulfils the criterion, the corresponding safety function is maintained) can be defined for specific safety functions.

The multi-barrier system approach has been applied to surface/near-surface and geological disposal concepts and designs. The wasteform itself often provides some degree of radionuclide retention. However, conservatively, no credit is usually taken for the retention of radionuclides by most wasteforms in deep disposal with the exception of vitrified high-level waste (HLW) and spent fuel (SF). For intermediate-level waste (ILW), depending on the disposal concept and host geology, their contribution may often be assumed to be negligible in comparison to components of the engineered barrier system (EBS) and to the geological host formation (which may be the main barrier to radionuclide migration). The main processes considered for IL-wasteforms are those related to their compatibility with EBS components and the host rock formation and any disturbances that might eventually be induced from these interactions. Typical examples include the release of complexing agents, gas generation and the effect of alkaline pore water migrating from wasteforms and the EBS into the surrounding rock (this is not an exhaustive list and depends on the disposal concept).

A similar conservative assumption about the contribution of the wasteform to radionuclide retention may also be made in surface disposal. In Belgium, for example, up to now the contribution of the wasteforms is not taken into account in the performance assessment, even if they would provide some degree of radionuclide retention in comparison to the other EBS components in this case (Wacquier et al. 2013). The reason is the great variety of wasteforms and the related complexity of the processes involved. Because of the remaining uncertainties, it would be difficult to change the conceptual model from something that is conservative and defensible (such as instant release of all the radionuclides from the wasteform) to more representative conceptual models that account for known physical and chemical processes. Adding model complexity adds to data requirements which have to be underpinned by extensive research and development (R&D) programmes. These added costs must be balanced against the benefits obtained for the entire safety assessment. Nevertheless in the future, for some particular cases, the contribution of a wasteform to radionuclide retention could possibly be considered if justified by a potential significant contribution to limiting the release of radionuclides and by a predictable behaviour. This could apply particularly to activated metals such as stainless and carbon steels where the ranges of the uniform corrosion rates of these alloys in cementitious environments have been determined with a good level of confidence.

Even if the contribution of the wasteform is not taken into account in performance assessments, their behaviour is the subject of R&D programmes. A minimum knowledge of their performance and expected evolution builds confidence (developing the scientific assessment basis), even if significant interactions with components of the EBS and with the host rock are not expected, and can quantify the safety reserve associated with them.

5.1.1 Performance Assessment

A performance assessment (PA) or safety assessment (a term used internationally) quantitatively estimates the potential post-closure impacts to human health associated with a radioactive waste facility. PAs are a means of helping decision makers to evaluate siting, design, operation and decommissioning of the waste disposal facility. PAs also identify a baseline point of compliance, require a sensitivity/uncertainty analysis and address requirements related to the protection of water resources. In some cases additional analyses are performed to identify doses to the public, not only from the disposal facility under consideration but from any other co-located sources that could contribute to a composite dose to a member of the public. These composite analyses are used to ensure that the total dose associated with the facility and any other source remains within levels allowed for exposure to the public. PAs of disposal systems are iterative processes involving site-specific modelling evaluations. The primary goal of a PA is to determine with reasonable assurance that the facility complies with regulatory criteria for the protection of human health and safety. The other goal of PAs is to identify critical data needed to support facility design and information needs. These are needed to make defensible and cost effective licensing decisions, and to maintain operating limits (e.g. waste acceptance criteria). The modelling conducted for PAs includes assessments of contaminant migration through environmental pathways (e.g. air, groundwater and surface water) and potential human exposures to the contaminants in various exposure media (e.g. soil, drinking water, crops and livestock). The potential for inadvertent human intrusion into the waste as an accidental pathway can also be modelled.

Figure 5.1 is a graphical representation of a PA process where the different components of the PA are represented as a function of time, infiltration, location and the chemical environment (adapted from US Nuclear Regulatory Commission 2000).

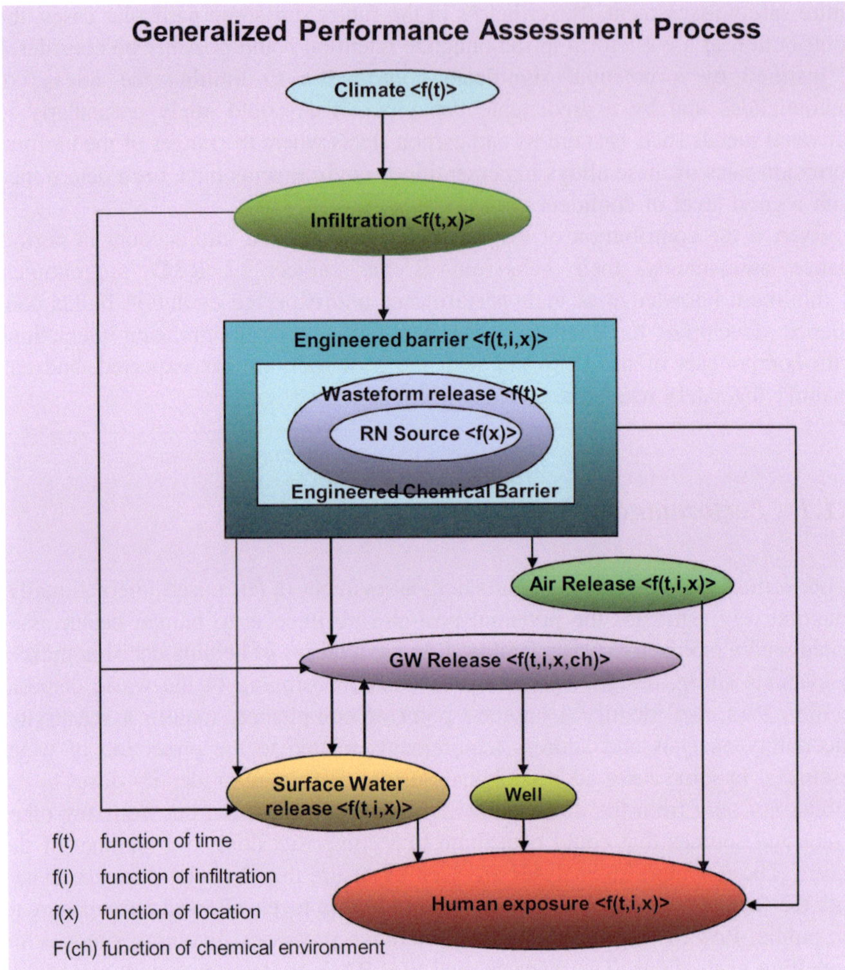

Fig. 5.1 Performance Assessment (PA) process

The central attribute of the process is that it is conducted iteratively, starting with a combination of generic and limited site-specific information in support of relatively simple conservative models and analyses, and progressing to more realistic, site-specific and detailed analyses, as necessary, to reduce uncertainty in assessing the performance of a disposal facility.

In current PA practice, in addition to the concept of an iterative process, PA is performed on a regular basis as information and data are acquired and as the performance assessment of a facility evolves. In the broader sense, PA has become a management tool in addition to a demonstration of compliance with regulatory criteria. This has led to increasingly sophisticated methods capable of representing physicochemical processes that can be used to guide improved designs for waste-forms, containments and facilities. As discussed later, there has been an increase in the use of stochastic modelling approaches to better capture the uncertainty in modelling results. Benefits of these improvements have been recognised in increased defensibility with stakeholders and improved efficiency of waste management practices.

As an example, in ONDRAF/NIRAS' proposed safety assessment methodology (ONDRAF/NIRAS 2013), safety assessments are carried out in two main phases at every programme stage (see Fig. 5.2):

- a phase of preparatory assessments
- a phase of formal assessments

Preparatory safety assessments are conducted continuously and repeatedly, at every programme stage, on the basis of phenomenological evidence from the

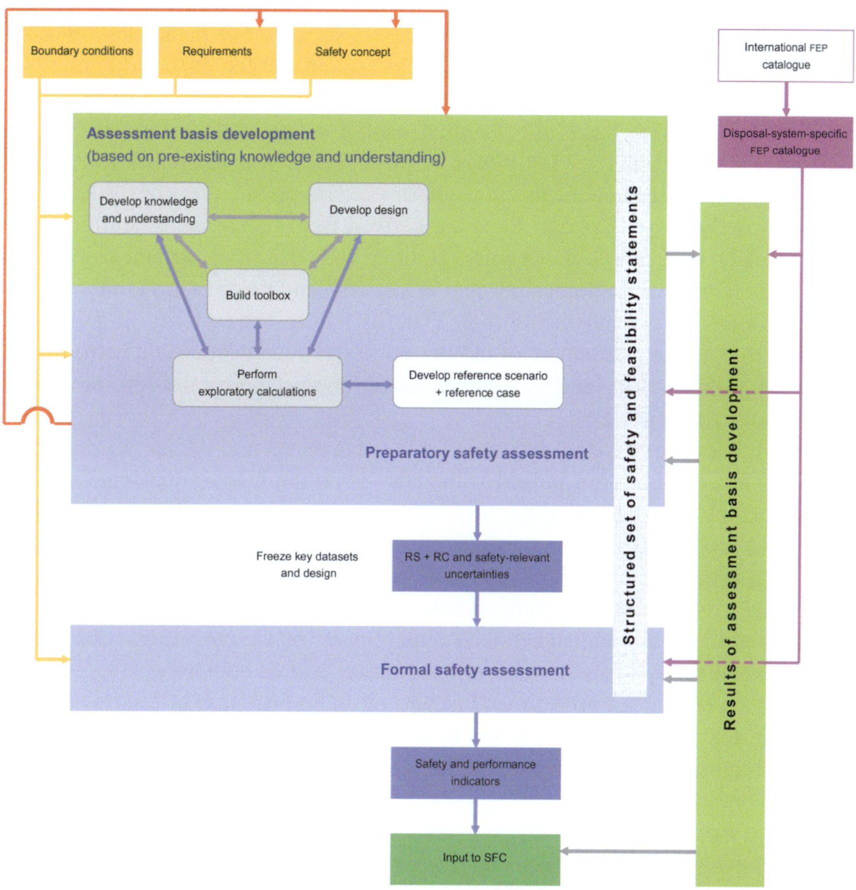

Fig. 5.2 ONDRAF/NIRAS proposed assessment methodology

assessment basis, and in parallel with system development. They take the form of sensitivity analyses and involve qualitative arguments. They thus entail sustained, structured interactions between safety assessors and experts in phenomenology and technology. Exploratory calculations are used, for example, to identify those contaminants that are potentially safety relevant, and therefore need to be considered in formal safety assessments, or to evaluate the impact of a particular process or uncertainty on system evolution and performance and hence determine its relevance to safety. Preparatory assessments also aim to identify any significant deficiencies in current knowledge and understanding and in the plans to address these in a R&D programme, thus providing feedback to assessment basis development. Modifications can then be made to the programme and, should this appear necessary, in the strategic choices (e.g. the choice of concept).

Through preparatory assessments, the safety analysts gain stepwise confirmation that the disposal system will evolve as defined in the safety concept and will fulfil each safety function as required (the reference scenario). They are then able to derive a reference case (i.e. a specific realisation of the reference scenario associated with conservative or a more realistic, but well-justified set of parameter values) as well as gaining insight into the impact of uncertainty on the expected evolution.

A methodological tool that can be used to examine the impact of perturbing phenomena and associated uncertainties on the safety and performance of the system in a systematic way is through a set of 'safety and feasibility statements'. Safety and feasibility statements are assertions regarding the safety or the feasibility of a proposed disposal system. They are derived from requirements and organised hierarchically in a top-down manner, starting with the most general (high-level) statements (e.g. 'The geological system is known') and progressing to increasingly specific (lower-level) statements (e.g. 'The hydraulic conductivity is sufficiently characterised'). The assessment of the impact of the uncertainties and perturbations on the validity of a statement, relative to the understanding needed, provides a useful tool to steer R&D.

Once the level of support available for the statements and the knowledge of the impact of uncertainties are judged sufficient (given the programme stage at hand) for proceeding with formal safety assessment, the key datasets within the assessment basis and the design are formally frozen. A formal safety assessment aims to show in a formal, quantitative way why the disposal system under consideration, despite the existence of remaining uncertainties, can be judged to be safe, satisfying all relevant regulatory and stakeholder requirements of the programme objective at hand. Formal safety assessments are quantitative, and as comprehensive as necessary, and are conducted prior to, and as part of, the preparation of a safety case (or safety and feasibility case). They aim to illustrate through a selection of calculation cases and indicators, the robustness of the system towards low probability scenarios affecting one or several safety functions (altered scenarios), alternative evolutions which are equally probable but do not affect the safety concept of the system (alternative cases within the reference scenario) and also, in particular, quantification of the safety margin with respect to the reference case. Formal safety assessments demonstrate how the system fulfils the regulatory and stakeholder requirements and highlight and prioritise the main open issues left to be treated in the next programme stage. Regulatory requirements may be, and are, different in different countries.

As part of the quality assurance of safety assessments, completeness checks are conducted during both preparatory assessments and formal assessments using a disposal-system-specific catalogue of features, events and processes (FEPs) (e.g. ONDRAF/NIRAS 2013) based on international FEP lists (NEA/OECD 2000).

PAs that include analyses of cementitious barriers have traditionally been associated with waste disposal. Current challenging environmental remediation and decommissioning activities, and disposal of low level, intermediate and high-level waste require PA-like analyses that may benefit from understanding the performance of cementitious materials. These involve numerous applications in many different locations (e.g. concrete vault-based designs for low-level waste disposal, in situ decommissioning of nuclear reactors and remediation of contaminated sites and old burial grounds). Typically, the performance of engineered barriers, including cementitious barriers, can be divided into those: (1) based on hydrological effectiveness or physical containment to reduce water-to-waste or waste-to-water contact; (2) chemical effectiveness to limit radionuclide transport; or (3) a combination of the two. Figure 5.3 depicts the time frames for apportioning approximate credit years for reliance on engineered/chemical barriers and site characteristics for Low Level Radioactive Waste (NRC Regulations 2016).

Concrete degradation mechanisms (e.g. sulphate attack, rebar chloride corrosion, cracking, alkali aggregate reaction, carbonation) can cause degradation and/or failure of the barrier and corresponding contaminant release. Obtaining data and

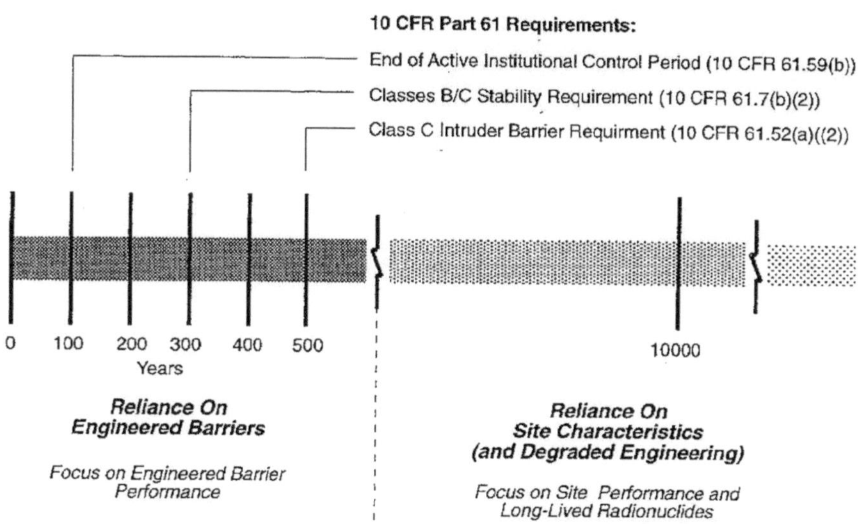

Fig. 5.3 Time frames for credits in PA for engineered barriers/site characteristics (taken from NRC Regulations 2016)

information from the cementitious barrier system facility is vital in the iterative PA process. This helps assure that the uncertainties in barrier performance are known and that any simplifying assumptions made for barrier performance in earlier PAs are still appropriate and, if not, revised. The revised PAs account for structural degradation for estimating whether end-of-life predictions in the PA are reasonable. Ideally, risk-based criteria are needed to evaluate how ageing and degradation affects structural capacity (structural margins) and whether PA accounts for it. For chemical containment, where the effectiveness of cementitious materials strongly depends on the source term, and source release characteristics, performance is very difficult to predict and is strongly related to both hydraulic properties and quantity of cement-based materials present. A cement-based barrier may also limit inadvertent intruder contact with waste, probably for at least a few hundred years, if it remains unexposed to aggressive environmental conditions. Because the performance of cementitious barriers may have to be assessed over hundreds if not thousands of years, the uncertainties for cementitious barriers are likely to be important to the PA.

5.1.2 Treatment of Uncertainties

The importance of sensitivity and uncertainty analyses has long been recognised as an integral part of PA for waste facility disposal. Figure 5.4 depicts an analysis process approach that incorporates parameter and model uncertainty in the PA calculations (US Nuclear Regulatory Commission 2000).

However there has not been general agreement regarding the specific approaches used to implement such sensitivity and uncertainty analyses. The views on sensitivity and uncertainty analyses can be different depending upon the regulatory environment, analyst preference and other reasons. For example PA for wasteforms from waste processing may have different goals than soil and groundwater assessment for remediation and may also be somewhat different from decommissioning assessments. Approaches to uncertainty analyses are an important consideration in the assessment of cementitious barrier performance in PA (Cementitious Barriers Partnership 2009).

Throughout the PA iterative process, sensitivity analyses are used to identify parameters with the greatest influence on the decision to be made and provide a means of focusing attention on those parameters for both the operations of the facility and compliance with regulations. For many years it was common to use deterministic approaches, which involved a base case and multiple sensitivity analyses targeted to explaining or better illustrating the effects of changes in different parameters on the overall result of the assessments over time. Lately, there has been increased use of probabilistic approaches to replace or supplement the deterministic calculation.

Two typical ways of classifying uncertainties are apparent in PA for health risk assessments. One method classifies uncertainties according to where in the risk

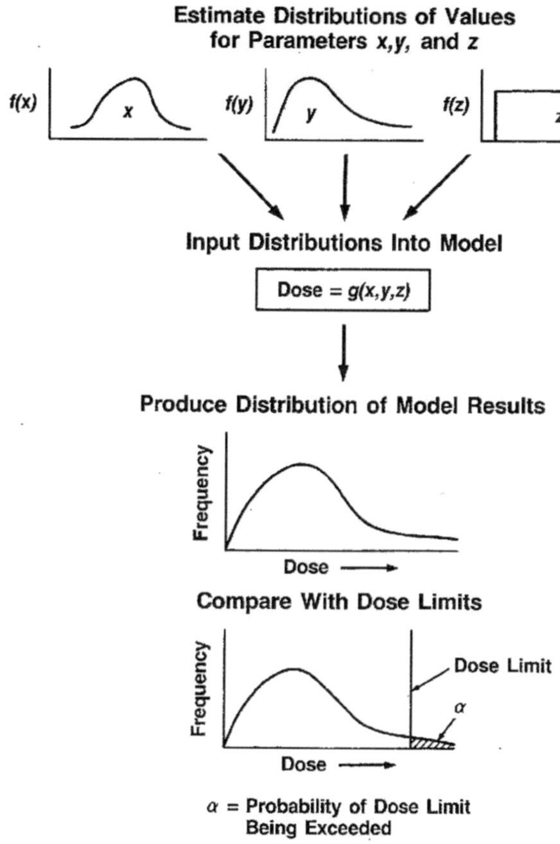

Fig. 5.4 Incorporation of model and parameter uncertainty in PA (taken from US Nuclear Regulatory Commission 2000)

assessment process they occur. Other approaches characterise uncertainties into more abstract, general categories, e.g. bias, randomness and variability. A more useful category is one based on parameter, model and scenario uncertainty. Parametric uncertainty occurs due to lack of knowledge of the 'true' value of an input parameter to a model. In model uncertainty, lack of knowledge about the structure and accuracy of the model is the major concern and includes simplifying assumptions and mathematical representations e.g. the use of 1-D and 2-D models. Scenario uncertainty arises from lack of information regarding missing or incomplete information needed to adequately define the model. This is an important aspect since PA of waste facilities involves predictions of performance, sometimes for hundreds to thousands of years, where the analysis of FEPs may play an important role. Overall, one element that runs through PAs and risk assessments is the need for expert judgment to be used to determine the appropriate parameters, values, distributions, models and scenarios. Expert judgment is valuable in that experts often have the greatest experience with these types of problems; however their judgment may suffer from the same biases as lay people. Use of a structured elicitation methodology is necessary in order to limit human bias (see, for example, United Kingdom Nirex Limited 2006).

It is important to consider the challenges specifically associated with the development of input distributions to support a probabilistic assessment. Multiple methods are available for characterising uncertainties in risk assessments. Two popular methods are Monte Carlo simulation and sensitivity analyses. For risk assessments Monte Carlo analyses involve characterising the uncertainty and variability in risk estimates by repeatedly sampling probability distributions representing risk equation inputs and using the results to estimate the range of risks. On the other hand sensitivity refers to variation in model inputs with respect to changes in model inputs that can provide a rank ordering of model inputs based on their relative contributions to model output variability and uncertainty. This can provide meaningful insight if the number of uncertain parameters is fairly small.

Characterising the properties and reducing uncertainties in understanding and predicting the fundamental behaviour of cementitious barriers is needed to evaluate and improve system design for near-surface engineered waste disposal systems (e.g., wasteforms, containment structures and entombments), environmental remediation and decommissioning. Uncertainty reduction can be informed by models coupling multi-scale and multi-physics processes, including physical chemical evolution and transport phenomena, and applying these to heterogeneous cementitious materials with changing boundary conditions. Ultimately, these can be integrated into a set of tools to predict the structural, hydraulic and chemical performance of cement based barriers over extended time frames [e.g. >100 years for operating facilities and >1000 years for nuclear waste management facilities (Cementitious Barriers Partnership 2009)].

5.2 Role and Durability of Cement Barriers in Waste Storage and Disposal

Zuloaga (2011) has discussed the ageing management program for the Spanish spent fuel and high level waste centralised storage facility. The planned operational life is 60 years, while the design service life is 100 years. Durability studies and surveillance of the behaviour have been considered from the initial design steps, taking into account the accessibility limitations and temperatures involved. The preliminary definition of the Ageing Management Plan, addressing the behaviour of spent fuel, its retrievability, the confinement system and the reinforced concrete structures, including test plans and surveillance design considerations, based on El Cabril LILW disposal facility developments, were presented.

Dry cask storage systems are used in the US for interim storage of spent fuel prior to the availability of a repository for these wastes in the future (Philip and

Graves 2013). Spent fuel is packaged within steel casks that are housed with an overpack, typically made from reinforced concrete, which acts as a storage module. These units are then placed vertically or horizontally on a concrete pad. These systems are currently licensed for 20 years with a potential for extension beyond this time. The concrete elements of these are subject to possible degradation through the effects of temperature and thermal gradients, ionising radiation, corrosive agents, water in air and soil, sulphate and chloride attack, alkali-silica reaction, carbonation and freeze-thaw cycling. A programme of modelling, inspection, monitoring, maintenance and repair of this type of container is required for successful ageing management. Snyder et al. (2006) discuss the application of Kalman filtering to combine computer model predictions with periodic measurements. An example scenario was presented based on the possible degradation of a concrete structure resulting from a concentration of a particular mobile species.

The impact of heat on the durability of concrete structures with regard to the long-term interim storage has been investigated by CEA (Lagrave et al. 2006). The technical approach is based on three areas of research: material characterisation; modelling to identify weaknesses in the structure; and validation by experimental tests on heavily instrumented structures subjected to representative loads. Models to estimate the effect of drying on the mechanical and the thermo-hydro-mechanical behaviour of concrete had been validated at the centimetre scale and were underway at full-scale (e.g. the GALATEE programme).

The role and performance of engineered barriers in present and future facilities for waste management and disposal in Romania have been described (Fako et al. 2013). Short-lived low and intermediate level wastes are conditioned by grouting with ordinary Portland cement in carbon steel drums and have been placed in disposal galleries in the low-level radioactive waste repository (DNDR) at Băita Bihor since 1985. Filled galleries are then isolated by the construction of a masonry wall. Since 1996 the galleries have also been backfilled with bentonite powder. The facility is located in an old mine and the geology and hydrogeology are highly disturbed and complex, leading to challenges in guaranteeing the long-term performance and safety of the facility. Current work has the objectives of increasing radiological safety by improving the technology for filling the galleries, developing a closure plan that is compliant with national and international requirements, and evaluating the performance of the closure systems. The concrete elements of the disposal system are the waste encapsulation matrix, which is an important physical and chemical barrier to the migration of contaminants and where improved formulations will be developed, and the concrete floors and walls, which provide a lesser benefit for the long-term safety of the repository. Spent fuel is stored at the Cernavodă interim dry spent fuel storage facility that was commissioned in 2003. Three of a total of 27 storage modules have been constructed so far and the condition of these is monitored. A near-surface facility (DFDSMA-Saligny) for LILW-SL (short-lived low-level and intermediate-level wastes) generated during the operation, refurbishment and decommissioning of the Cernavodă NPP is planned. The concrete barriers of the multiple engineered barrier conceptual design would play a more significant role in the long-term safety of the facility compared to those in DNDR and it is intended to model the system evolution. The waste encapsulation grout will be finalised during implementation of the conceptual project.

Kari and Puttonen (2009) discuss finite element modelling of the durability of concrete engineered barriers in the context of the disposal concept for low- and intermediate-level waste in Finland where these barriers are required to achieve a service life of at least 500 years after the facilities are sealed. The modelling considered interactions of atmospheric carbonation, chloride penetration, concrete degradation due to sulphate and magnesium, and leaching. Preliminary modelling results were compared with results from experiments and empirical models. It was concluded that the interaction of deterioration mechanisms is important and that the long-term deterioration of reinforced concrete may not be assessed with sufficient accuracy if only a single phenomenon is considered. Further discussion of this modelling approach was presented at NUCPERF 2012 (Kari and Puttonen 2013). It is important to identify the most critical model parameters in probabilistic calculations to assess the durability of concrete structures. Gulikers (2006) discusses this in relation to chloride ingress and shows that the age factor and critical chloride content have a pronounced influence on service life predictions. He concluded that the choice of mean values for model parameters has to be objective and care should be exercised when using values from limited data sets; this may lead to the necessity of making conservative choices. Reliable field data from real structures with information on measured chloride profiles, exposure conditions and concrete quality are required to derive more reliable statistical quantification of model parameters.

5.3 Impact of Gas Generation and Carbonation of Cement Systems

Simulations of atmospheric carbonation of concrete intermediate-low level waste (ILLW) packages during an operating period of up to 100 years with ventilation have been undertaken (Thouvenot et al. 2013). These considered the ILLW disposal zone within a facility based in the deep Callovo-Oxfordian formation in France according to the concept developed by ANDRA. Ventilation will lead to the desaturation of cement and atmospheric carbonation, which could potentially trigger corrosion of steel reinforcement. Atmospheric carbonation of concrete in unsaturated conditions is a complex process involving coupling of transport of water and CO_2 in the gas and liquid phases, capillary flow during drying and chemical reactions involving cement hydrates and CO_2 dissolved in the liquid water phase. Initial modelling of the drying and carbonation of concrete using the ToughReact code calculated a carbonation depth of between 1 and 4 cm in 100 years, depending on the assumed properties of the concrete. The degree of carbonation was greater than that found in experiments and this was attributed to decreasing reactivity as the degree of saturation of concrete fell from around 0.7 to

0.3. The model was refined to include this decreasing reactivity, which resulted in less intense but deeper carbonation of the concrete. Thouvenot et al. consider that the model could be further refined by taking into account the protective effect of secondary minerals, which would reduce carbonation intensity. In addition, they suggest that a more realistic simulation of the alteration phases may be possible in the future if kinetic data for the alteration of CSH at low Ca:Si ratios become available.

Carbonation was also the degradation mechanism considered in a risk analysis of possible concrete degradation with respect to safety of a proposed near-surface disposal facility for short-lived low and intermediate-level waste in Belgium (Capra et al. 2013). Carbonation, which could induce corrosion of rebar in the waste package monoliths or the module walls and slabs, was considered over a period of 2000 years from the commencement of operation of the facility until the time at which chemical containment of radionuclides was no longer claimed in the safety case. The risk analysis focused on the safety function "limitation of water through the system (thanks to the engineered barrier)". The time evolution of the carbonation front in the concrete elements was assessed probabilistically for several combinations of temperature, humidity and CO_2 content. The probability of the carbonation front being deeper than the thickness of the concrete cover of the rebars represents the end of the initiation period and the possible start of corrosion along the rebars. The impacts on the safety function were then ranked according to their probability, and their effect on the limitation of water flow for the component considered, in terms of localised, extended or global impact for the facility. It was found that there was no critical risk to limitation of water flow over a 2000 year period for any individual component. Minor risks were identified that could be managed by quality control during construction and the use of the specific concrete formulation proposed. Lower risks, after closure of the facility and the end of the nuclear regulatory control period after 350 years, related to the effects of inadvertent removal of the earth embankments or cover and could be managed by appropriate surveys of the disposal facility.

The Cementitious Barriers Partnership (Kosson 2013) has the aim of developing a set of tools to predict the structural, hydraulic and chemical performance of cement barriers used in nuclear applications over extended time frames (e.g. up to and greater than 100 years for operating facilities and greater than 1000 years for waste management) on behalf of the Office of Environmental Management in the US Department of Energy. A range of reference cases (e.g. cementitious wasteforms in a concrete disposal vault with a cap, grouted high-level waste tank closure and grouted vadose (unsaturated) zone contamination) have been identified for the comparison and demonstration of tools developed by the partnership. Codes such as STADIUM and LeachXS™/ORCHESTRA (http://cementbarriers.org/partner-codes) can be coupled to GoldSim (http://www.goldsim.com) in order to undertake probabilistic calculations and analyses of uncertainty and sensitivity. One important challenge that the Cementitious Barriers Partnership is currently addressing is the assessment of the integrity and closure of high-level waste tanks after they have been emptied and grouted internally (Brown et al. 2013). The HLW tanks may be empty for many years prior to closure and the performance of the closed tanks must be assessed over centuries, if not millennia. Carbonation-induced corrosion of the steel tanks (due to carbonation of the outer concrete shell) was identified as a primary degradation mechanism and possible failure mechanism prior to closure. After closure the impact of carbonation and concurrent oxidation of redox-sensitive elements may be to increase the release and short-range transport of some contaminants. LeachXS™/ORCHESTRA was used to model the penetration of $CO_2(g)$ into the uncracked partially saturated wall of a representative HLW tank, the resulting carbonation reaction and leaching of constituents. Parameters including composition, soil-gas CO_2 concentration, degree of concrete saturation, porosity, CO_2 effective diffusivity, the mineral assemblage and thermodynamic parameters were varied to evaluate the sensitivity of the calculated results. The variations in these parameters tended to have less than one order of magnitude effect on the predicted extent of carbonation resulting in a pH value of less than 9, at which embedded carbon steel can be depassivated and become liable to corrosion. The influence of carbonation and chloride ingress in reinforced concrete structures is also of relevance to the corrosion of rebars in concrete secondary containment vessels for nuclear power plants (Petre-Lazar et al. 2006).

The importance of water content, cement porosity, iron content and gamma dose in determining the evolution of hydrogen from pore water radiolysis by βγ-emitting wastes encapsulated in cement binder was illustrated by the results of a study using the model DOREMI (Bouniol and Bjergbakke 2008) undertaken by Foct et al. (2013). This showed that the radiolytic production of molecular hydrogen depends strongly on water saturation with a maximum production rate occurring near 64 % saturation. Hydrogen production was calculated to decrease sharply at higher saturations due to the decrease in open porosity available for gaseous diffusion and a resultant increase in hydrogen recycling. The results also showed that increased cement binder porosity increased hydrogen production (because of the increased water content), as did increased dose rate. The presence of iron at greater than 64 % water saturation also increased radiolytic dihydrogen production because of the preferential reaction of Fe(II), rather than H_2, with the $O^{\cdot-}$ radical.

5.4 Conclusions

In summary, given that cementitious materials are engineered features, the key uncertainties in performance assessments tend to be related to the size of radioactive source term and near-field transport. The amount of credit taken in PAs for the role of cementitious barriers ranges from taking no credit at all, through to taking considerable credit for the physical and chemical properties (including the timing of any degradation). In some cases simplifying conservative assumptions may be made to take account of uncertainty when cementitious materials are considered in PAs. Such assumptions are generally made because of a lack of site- and facility-specific information for the cementitious materials or for expediency where conservative assumptions may be shown to be adequate. Alternatively, the uncertainty may be quantified and sampled in a probabilistic model.

In general, the role of cementitious materials as a physical barrier is for shorter periods than for their role as a chemical barrier. Cementitious barriers are designed to physically contain short-lived radionuclides, whereas their chemical properties limit radionuclide release rates for many longer-lived radionuclides. Because of the difficulties associated with quantifying the extent and initiation of cracking of cementitious materials when assessing the physical containment function, simple assumptions are often made (e.g. the cementitious barriers fail completely at the onset of through-wall cracking). For chemical barriers, the most common consideration has been the use of empirical sorption distribution coefficients (K_d values) that account for the radionuclide retardation properties of cementitious materials. The presence of reducing conditions in grouted wastes is an important consideration when considering redox-sensitive radionuclides (e.g. ^{99}Tc) in PA. This approach can lead to improved confidence related to taking credit for the long-term chemical performance as opposed to assessing the evolution of concrete cracking over time and its impact on physical containment.

5.5 RILEM TC-226-CNM Recommendations for Future R&D

There are a number of areas where improved understanding would be of benefit in underpinning the assessment of the performance of cementitious barriers in radioactive waste management.

Because of the difficulties associated with quantifying the extent and initiation of cracking of cementitious materials when assessing physical containment, simplifying assumptions are often made, for example that the cementitious barriers fail completely at the onset of through-wall cracking. A better understanding of how cracking evolves and how this leads to a developing loss of physical containment would allow credit to be taken for more gradual changes as cracking progresses.

The aqueous concentration of many radionuclides may be limited by the formation of low-solubility solid phases and through sorption to cement surfaces under the aqueous conditions of high pH imposed by the cementitious materials and low redox potential developed through components of the wasteform or engineered barrier system (e.g. metal corrosion, grouts containing BFS). The possible spatial and temporal evolution of these conditions is important when considering the range or uncertainty of the radionuclide parameters to be used in performance assessments. Research is ongoing to give increased understanding of how such conditions evolve (e.g. through improved thermodynamic and physical data and modelling of new phases) and is to be encouraged.

Carbonation affects the ability of cement materials to condition pore water to high pH. This can lead to increased rates of corrosion of embedded steels (e.g. reinforcement and closed underground tanks) and increased radionuclide migration.

Much modelling of carbonation has been carried out under partially-saturated conditions and further refinement to such models can be expected if data on secondary minerals and kinetics of carbonation of low Ca:Si gels are obtained. The rate and effect of carbonation of cements under fully-saturated conditions has been studied less extensively and such studies would be relevant to deep disposal facilities.

Evaluations of the performance of cement structures should include an appropriate sensitivity analysis to take account of variability in operations and materials. There is a general need for the development of probabilistic approaches to account for uncertainty in materials characteristics and performance. Overall the evolution of cement barriers should be considered in the context of the safety functions they are required to provide in a particular waste management concept.

References

Bouniol, P., Bjergbakke, E.: A comprehensive model to describe radiolytic processes in cement medium. J. Nucl. Mater. **372**, 1–15 (2008)

Brown, K.G., Arnold, J., Sarkar, S., Flach, G., van der Sloot, H., Meeussen, J.C.L., Kosson, D.: Modeling carbonation of high-level waste tank integrity and closure. In: Proceedings of NUCPERF 2012 Workshop, Long-Term Performance of Cementitious Barriers and Reinforced Concrete in Nuclear Power Plant and Radioactive Waste Storage and Disposal, Special issue of The European Physical Journal, vol. 56 (2013)

Capra, B., Billard, Y., Wacquier, W., Gens, R.: Risk assessment associated to possible concrete degradation of a near surface disposal facility. In: Proceedings of NUCPERF 2012 Workshop, Long-Term Performance of Cementitious Barriers and Reinforced Concrete in Nuclear Power Plant and Radioactive Waste Storage and Disposal, Special issue of The European Physical Journal, vol. 56 (2013)

Cementitious Barriers Partnership: US DOE and US NRC performance assessment approaches. CBP-TR-2009-001, May 2009, http://cementbarriers.org (2009)

NRC Regulations, Title 10, Code of Federal Regulations, Chapter1, Part 61 Licensing requirements for land disposal of radioactive waste. U.S. Nuclear Regulatory Commission, Washington, DC. reviewed/updated June 23, 2016, http://www.nrc.gov/reading-rm/doc-collections/cfr/part061 (2016)

Esh, D.W., Philip, J., Snyder, K.A.: Performance assessments insights on the use of cements in waste management. In: AMP 2010, International Workshop on Ageing Management of Nuclear Power Plants and Waste Disposal Structures (EFC Event 334), The European Physical Journal, vol. 12 (2011)

Fako, R., Barariu, G., Toma, R., Geotgescu, R., Sociu, F.: Analyze of intended long-term performances of cementitious engineered barriers for future storage and disposal facilities for radioactive wastes in Romania. In: Proceedings of NUCPERF 2012 Workshop, Long-Term Performance of Cementitious Barriers and Reinforced Concrete in Nuclear Power Plant and Radioactive Waste Storage and Disposal, Special issue of The European Physical Journal, vol. 56 (2013)

Foct, F., Di Giandomenico, M.-V., Bouniol, P.: Hydrogen production caused by pore water radiolysis in cemented intermediate level waste containers. In: Proceedings of NUCPERF 2012 Workshop, Long-Term Performance of Cementitious Barriers and Reinforced Concrete in Nuclear Power Plant and Radioactive Waste Storage and Disposal, Special issue of The European Physical Journal, vol. 56 (2013)

Gulikers, J.: Considerations on the reliability of service life predictions using a probabilistic approach. J. Phys. IV France **136**, 233–241 (2006)

Kari, O.-P.J., Puttonen, J.A.: Modelling the durability of concrete for nuclear waste disposal facilities. In: L'Hostis, V., Gens, R., Gallé, C. (eds.) Proceedings of the RILEM Workshop on Long-Term Performance of Cementitious Barriers and Reinforced Concrete in Nuclear Power Plants and Waste Management, EFC Event 317, RILEM Proceedings PRO 64, ISBN: 978-2-31158-072-pp. 167–174 (2009)

Kari, O.-P.J., Puttonen, J.A.: Comprehensive model for ageing of reinforced concrete structures under final disposal conditions. In: Proceedings of NUCPERF 2012 Workshop, Long-Term Performance of Cementitious Barriers and Reinforced Concrete in Nuclear Power Plant and Radioactive Waste Storage and Disposal, Special issue of The European Physical Journal, vol. 56 (2013)

Kosson, D.: Cementitious barriers partnership—project overview and toolbox release. In: Proceedings of NUCPERF 2012 Workshop, Long-Term Performance of Cementitious Barriers and Reinforced Concrete in Nuclear Power Plant and Radioactive Waste Storage and Disposal, Special issue of The European Physical Journal, vol. 56 (2013)

Lagrave, H., Ranc, G., Gallé, C., Durand, S.: Durability design of heated concrete structures, methodology and application to long-term interim storage. J. Phys. IV France **136**, 223–231 (2006)

NEA/OECD: Features, events and processes (FEPs) for geologic disposal of radioactive waste, an international database (2000)

Nuclear Decommissioning Authority: Geological disposal: near-field evolution status report. NDA Report NDA/RWMD/033 (2010)

ONDRAF/NIRAS: Research, development and demonstration (RD&D) plan for the geological disposal of high-level and/or long-lived radioactive waste including irradiated fuel if considered as waste, State-of-the-art report as of December 2012, ONDRAF/NIRAS Report NIROND-TR 2013-12 E (2013)

Petre-Lazar, I., Sudret, B., Foct, F., Granger, L., Dechaux, L.: Service life estimation of the secondary containment with respect to rebar corrosion. J. Phys. IV France **136**, 201–212 (2006)

Philip, J., Graves, H,: Long-term performance of dry cask storage systems (DCSS). In: Proceedings of NUCPERF 2012 Workshop, Long-Term Performance of Cementitious Barriers and Reinforced Concrete in Nuclear Power Plant and Radioactive Waste Storage and Disposal, Special issue of The European Physical Journal, vol. 56 (2013)

Snyder, K.A., Lu, Z.Q., Philip, J.: Long-term monitoring strategy for concrete-based structures using nonlinear Kalman filtering. J. Phys. IV France **136**, 191–200 (2006)

Thouvenot, P., Bildstein, O., Munier, I., Cochepin, B.: Numerical simulations of atmospheric carbonation in concrete components of a deep geological intermediate-low level waste disposal. In: Proceedings of NUCPERF 2012 Workshop, Long-Term Performance of Cementitious Barriers and Reinforced Concrete in Nuclear Power Plant and Radioactive Waste Storage and Disposal, Special issue of The European Physical Journal, vol. 56 (2013)

United Kingdom Nirex Limited: A procedure for data elicitation in support of performance assessments, Nirex report N/132 (2006)

US Nuclear Regulatory Commission: A performance assessment methodology for low-level radioactive waste disposal facilities—recommendations of NRC's performance assessment working group, US Nuclear Regulatory Commission report NUREG-1573, Washington DC (2000)

Wacquier, W., Cool, W., Bastiaens, W., Van Cotthem, A.: The design methodology and the safety concept of the surface disposal facility for the Belgian category A waste at Dessel. In: Proceedings of the ASME 2013 15th International Conference on Environmental Remediation and Radioactive Waste Management, ICEM2013, September 8–12, 2013, Brussels, Belgium (2013)

Zuloaga, P.: Ageing management program for the spanish spent fuel and high level waste storage facility. In: AMP 2010, International Workshop on Ageing Management of Nuclear Power Plants and Waste Disposal Structures (EFC Event 334), The European Physical Journal, vol. 12 (2011)

Chapter 6
Conclusions

Valérie L'Hostis and Robert Gens

Abstract After the three workshops organised and the edited proceedings produced: NUCPERF 2009 and 2012, and AMP 2010, the TC identified some R&D drivers. These fields of interest for R&D are presented in the present chapter.

Keyword R&D drivers · Processes · Properties · Understanding

The use of reinforced concrete in modern applications dates back around 100 years. The long-term service life and performance of concrete may be predicted with some confidence. For disposal systems, the longer time periods for performance necessitate confidence building rather than demonstration. Confidence may be supported not only by modelling but by showing robust solutions, together with a good knowledge of the processes involved. Confidence building is a continuous and iterative effort. Continuous improvement of the knowledge is the basis for providing confidence and supporting improvement of the safety assessment. The assessment of the safety of a disposal system requires a detailed knowledge base and adequate analysis tools: these form the assessment basis, whose development, through an iterative and stepwise process, is a major objective of the R&D programme.

The key degradation processes which will determine the long-term behaviour of EBSs should be identified for specific contexts and environments. A prerequisite for assessing the subsequent evolution of the system is good knowledge and understanding of its initial state.

Phenomenological studies include laboratory tests, in situ site characterisation studies, full-scale tests and demonstrations, and studies of natural and anthropogenic analogues. Experts in phenomenology consider the evolution of the

V. L'Hostis (✉)
CEA, Saclay, France
e-mail: valerie.lhostis@cea.fr

R. Gens (✉)
ONDRAF-NIRAS, Brussels, Belgium
e-mail: r.gens@nirond.be

V. L'Hostis and R. Gens (eds.), *Performance Assessment of Concrete Structures and Engineered Barriers for Nuclear Applications*,
RILEM State-of-the-Art Reports 21, DOI 10.1007/978-94-024-0904-8_6

disposal system including all its interactions (thermal, hydraulic, mechanical, biological and chemical) and pursue the goal of continuously refining their understanding of the spatial and temporal evolution of the system through an iterative R&D programme.

Analyses of natural and archaeological analogues such as Roman concretes are a good example of verifying the stability of hydrated phases that can be expected in thousands of years. It has to be acknowledged that a lot of work has already been done but key issues still have to be addressed and/or their understanding deepened in order to build confidence.

The R&D drivers identified by the Technical Committee (TC) are:

- New materials now being considered
- Extension of facilities lifetime (NPP)
- Performance assessment capabilities

The TC identified a number of fields of interest for R&D focus (not an exhaustive list). It is important to stress that it is not necessarily true that because a subject has been identified by the TC it is relevant in all cases, for all systems. Their relevance is case dependent (to be assessed case by case taking into account the specific details of each application).

We briefly list here some key issues:

- Multiple processes and properties have to be considered as a function of material composition
- Need for development of probabilistic approaches to account for uncertainty in materials characteristics and performance
- Models need to take account of experimental data on real systems
- Need for guidance to establish consistent and complete data sets
- Need to understand ageing and degradation phenomena for new materials (e.g., carbon fiber-cement paste interfaces in bulk material and exposure face, low pH cement)
- Potential consequences of passive corrosion on the long-term integrity of concrete (cracking)
- Need to couple corrosion, radiation and geochemical models
- Need for appropriate instrumentation/sensors to determine the ageing of the structures as well as to provide information for the proper design of future facilities
- Development of methodologies for inspection of inaccessible areas such as emergency cooling water tanks, grouted tendons in pre-stressed systems, etc.
- Effective ageing management for concrete structures with alkali-aggregates-reaction and delayed ettringite formation
- Need for a better understanding of how cracking evolves and how this leads to a developing loss of physical containment
- Potential influence of spatial and temporal evolution of geochemical conditions on radionuclide parameters to be used in performance assessments
- Impact of carbonation on corrosion and radionuclides transport